Mediterranean Mountain Environments

Mediterranean Mountain Environments

Edited by

Ioannis N. Vogiatzakis

School of Pure and Applied Sciences, Open University of Cyprus, Nicosia, Cyprus

WILEY-BLACKWELL

A John Wiley & Sons, Ltd., Publication

This edition first published 2012
© 2012 by John Wiley & Sons, Ltd

Wiley-Blackwell is an imprint of John Wiley & Sons, formed by the merger of Wiley's global Scientific, Technical and Medical business with Blackwell Publishing.

Registered Office: John Wiley & Sons, Ltd, The Atrium, Southern Gate, Chichester, West Sussex, PO19 8SQ, UK

Editorial offices: 9600 Garsington Road, Oxford, OX4 2DQ, UK
The Atrium, Southern Gate, Chichester, West Sussex, PO19 8SQ, UK
111 River Street, Hoboken, NJ 07030-5774, USA

For details of our global editorial offices, for customer services and for information about how to apply for permission to reuse the copyright material in this book please see our website at www.wiley.com/wiley-blackwell.

The right of the author to be identified as the author of this work has been asserted in accordance with the UK Copyright, Designs and Patents Act 1988.

Library of Congress Cataloging-in-Publication Data

CIP data has been applied for.

A catalogue record for this book is available from the British Library.

Wiley also publishes its books in a variety of electronic formats. Some content that appears in print may not be available in electronic books.

Set in 10.5/12.5pt Times by Aptara Inc., New Delhi, India.
Printed and bound in Singapore by Markono Print Media Pte Ltd.

First Impression   2012

To Isabella,
who will soon start enjoying the Mediterranean and its mountains.

Contents

5 Climate and hydrology   87
Carmen de Jong, İbrahim Gürer, Alon Rimmer, Amin Shaban and Mark Williams

6 Biogeography   115
Ioannis N. Vogiatzakis

7 Cultural geographies   137
Veronica della Dora and Theano S. Terkenli

8 Land use changes   159
Vasilios P. Papanastasis

List of contributors

Veronica della Dora   School of Geographical Sciences, University of Bristol, University Road, Clifton, Bristol, BS8 1SS, UK

İbrahim Gürer   Gazi University, Mühendislik Fakültesi Eti Mah. Yükseliş Sok No:5 Maltepe Ankara, Turkey

Philip D. Hughes   Geography, School of Environment and Development, Arthur Lewis Building, University of Manchester, Manchester, M13 9PL, UK

Carmen de Jong   The Mountain Centre, University of Savoy, F 77376 Le Bourget-du-Lac, France

Stefano Loddo   AGRIS (Sardinian Agency for Agricultural Research), Viale Trieste 111, 09123 Cagliari

J.I. López-Moreno   Pyrenean Institute of Ecology, CSIC, Spain

A.M. Mannion   Department of Geography and Environmental Science, School of Human and Environmental Sciences, The University of Reading, Whiteknights, PO Box 227, Reading, RG6 6AB, UK

Maria Teresa Melis   Laboratory of Photogeology and Remote Sensing, Department of Earth Sciences, University of Cagliari, Via Università 40, 09124 Cagliari, Italy

David Nogués-Bravo   Center for Macroecology, Evolution and Climate, University of Copenhagen, Denmark

Vasilios P. Papanastasis   Laboratory of Rangeland Ecology, Aristotle University Campus (286), School of Forestry and Natural Environment, 54124, Thessaloniki, Greece

Alon Rimmer   Israel Oceanographic & Limnological Research Ltd, The Kinneret Limnological Laboratory, PO Box 447, Migdal 14950 Israel

Amin Shaban   National Council for Scientific Research, CNRS, 59, Zahia Salmane street, Jnah, P.O. Box 11-8281, Beirut, Lebanon

Theano S. Terkenli   Department of Geography, University of the Aegean, University Hill, 81100 Mytilene, Greece

S.M. Vicente-Serrano   Instituto Pirenaico de Ecología, CSIC, Zaragoza, Spain

Ioannis N. Vogiatzakis   School of Pure and Applied Sciences, Open University of Cyprus, PO Box 12794, 2252 Latsia, Nicosia, Cyprus

Mark Williams   Department of Geography and Institute of Arctic and Alpine Research (INSTAAR), 450 CB, Boulder Colorado 80309-0450 USA

Preface

After the publication in 1992 of J.R. McNeil's *The Mountains of the Mediterranean World* and despite the general textbooks on the physical geography of the Mediterranean, there has not been a book dedicated to the mountains of the region. There is currently increasing interest for teaching and research on the topic due to rapid changes induced by anthropogenic forces, including changes in primary economic sectors operating in those areas (such as agriculture, forestry, etc.) and potential impacts of climate change. Current shifts in people's perception about the countryside have increased tourism and recreation pressures on both the natural and cultural environment, posing new threats but also providing new opportunities for the sustainable development of the Mediterranean mountains. The importance of interdisciplinary approaches to the study of mountain environments has recently been highlighted by both academics and practitioners; however, there is still a lack of integrated reading material on Mediterranean mountains that encapsulates both natural and cultural elements of the mountain environments in the region.

Mediterranean mountains unsurprisingly share many common biotic ecological, physical and environmental elements with mountains worldwide. At the same time they also display specificities as a result of a long and complex human presence in the area. However, they do not receive the same attention as their American, Asian and European counterparts.

Compared to the environmental historical account by McNeil, this volume is more of a textbook with chapters on physical and human geography of the mountains that correspond to topics taught in undergraduate courses at various universities worldwide. It is therefore much hoped that it will be received and adopted as such. This is hopefully just the start that will trigger a greater response to Mediterranean mountains, their landscapes and people.

Ioannis N. Vogiatzakis

Acknowledgements

The following colleagues, in alphabetical order, have kindly provided reviews for the chapters of this book: Dr Elias Dimitriou, Dr Rosario Gavilán, Dr Geoffrey Griffiths, Dr Steve Gurney, Prof. Jala Makhzoumi, Dr Jeroen Schoerl, Prof Jean-Pierre Suc, Dr Maria Zomeni. Dr Zomeni also assisted with proof reading while Māra Zēna and Vassilis Trigkas assisted with map production.

Chapter 5 is based on the MountSnowMed project proposal coordinated by the Prof. Carmen De Jong with a consortium of 29 partners from 13 countries around the Mediterranean in preparation for a 7th framework call of the EU in 2007. Our particular thanks go to Pier Carlo Zingari, Richard Escadafal, Gilles Neveau, Sergio Vicente-Serrano, Javier Sanchez Gutierrez, Rut Aspizua Canton, Vincente Carabias-Huetter, Louis Gimeno, Marco Borga, Spassimir Tonkov, Khier Bouguerra, Iggy Litaor, Mario Gomarasca, Alberto Bellin, Reinhard Böhm, Mehmet Oezel, Hulya Yildirim, Catherine Freissinet, Daniel Viviroli, Nektarios Chrysoulakis, Anna Spiteri, Hilmi Salem, Dan Blumberg, Dan Malkinson, Ivanka Dimitrova, Bruno Maiolini, Jacques Ganoulis for contributing to the rich reflections in the projects proposal. We are very grateful to the French Ministry of National Education, Higher Education and Research, Direction of Research and Innovation for financing the preparation of the proposal with their subvention Nr. 07A 706 in 2007. This enabled the organisation of an international partner meeting in Chambery, France, an extensive literature review, collection of basic data and creation of maps. We would also like to thank the input from Mark Williams and Pierre Paccard in preparing the project proposal.

Chapter 9 has been supported by funding from the Danish National Research Foundation, the research projects CGL2011-27574-CO2-02 and CGL2011-27536 financed by the Spanish Commission of Science and Technology and FEDER, ACQWA (FP7-ENV-2007-1- 212250) financed by the 7th Framework Programme of the European Commission, "*Efecto de los escenarios de cambio climático sobre la hidrología superficial y la gestión de embalses del Pirineo Aragonés*" financed by "Obra Social La Caixa" and the Aragón Government and *Influencia del cambio climático en el turismo de nieve* CTTP01/10, Financed by the *Comisión de Trabajo de los Pirineos*.

Permission to use a number of figures in this book were kindly granted by the Israel Sciences Journals – L.P.P. Ltd, Wiley, Elsevier, Springer, Oxford University Press, the British Library and the Gennadius Library at the American School of Classical Studies in Athens. At Wiley-Blackwell I would like to thank Fiona Seymour for her support throughout this endeavor, the production team and the copy editor Robert Hine for his professionalism.

Last but by no means least, I would like to express my gratitude to Laura [*grazie di cuore*] who has shown yet more patience with my new venture.

1

Introduction to the Mediterranean mountain environments

Ioannis N. Vogiatzakis

1.1 Introduction

Mountains are present in all continents, latitude zones and principal biome types, accounting for more than 20% of the Earth's terrestrial surface (Beniston, 2000). They come in all shapes and forms and are even present on islands, oceanic and continental. The northern hemisphere hosts most of the world's mountain areas, whereas the highest concentration of high mountains is in Central and southern Asia. The harsh conditions of mountain environments, including high altitude steep slopes, and extreme weather, result in them being regarded as hostile regions and therefore less inhabited and productive areas. However, they are still home to 20% (1.2 billion) of the world's human population and have special spiritual, cultural and sacred significance for over one billion people worldwide (Price, 2004). Isolation in geological and historic times has resulted in mountains acting as biological and cultural laboratories.

Worldwide mountains encompass a great diversity of topographic, climatic, biotic and cultural elements and therefore provide a range of ecosystem services (MEA, 2005).

> Mountains are an important source of water, energy and biological diversity. Furthermore, they are a source of such key resources as minerals, forest products and agricultural products and of recreation. As a major ecosystem representing the complex and interrelated ecology of our planet, mountain environments are essential to the survival of the global ecosystem.
>
> Agenda 21, Chapter 13, 'Managing Fragile Ecosystems:
> Sustainable Mountain Development'

Mediterranean Mountain Environments, First Edition. Edited by Ioannis N. Vogiatzakis.
© 2012 John Wiley & Sons, Ltd. Published 2012 by John Wiley & Sons, Ltd.

Half of the human population depends on mountains, while globally about half of the mountain area is under some sort of human land use (Körner and Ohsawa, 2005). Despite the harsh environmental conditions, human presence in mountain areas has a long history, going back millennia for some parts of the world such as the Mediterranean Basin. The wealth of goods and services that mountains provide come at an extra cost for human communities due to the limitations on the exploitation of these resources compared to other environments. The major environmental issues that mountains face worldwide include, other than dynamic geophysical processes (e.g. North Atlantic Oscillation and volcanic activity), anthropogenic ones such as pollution, land use change and human-induced climatic change (Beniston, 2000). A recognition of mountains in the policy agenda came in 1992 at the UN Conference on Environment and Development, which resulted in the establishment of other initiatives – Mountain Partnership, Mountain Forum: Glochamore (Global Change in Mountain Regions) – and received the attention of the Intergovernmental Panel on Climate Change, the Convention on Biological Diversity and the Millennium Ecosystem Assessment (Price, 2007). These are attempts to outline the scientific challenges (i.e. threats and opportunities) that the mountains face worldwide for future development (Table 1.1). Globally there are many drivers that affect

Table 1.1 Research goals for mountains according to the Glochamore initiative (Björnsen Gurung, 2005)

Topic	Research goal
Climate	To develop consistent and comparable regional climate scenarios for mountain regions, with a focus on Mountain Biosphere Reserves
Land use change	To monitor land use change in mountain regions using methods that are consistent and comparable
Cryosphere	To predict the areal extent of glaciers under different climate scenarios
Water systems	To determine and predict water balance and its components, particularly run-off and water yield of mountain catchments (including wetlands and glaciers) under different global change scenarios
Ecosystem functions and services	To predict the amount of carbon and the potential yield of timber and fuel sequestered in forests under different climate and land use scenarios
Biodiversity	To assess current biodiversity and to assess biodiversity changes
Hazards	To predict changes of lake systems and incidence of extreme flows in terms of frequencies and amounts, under different climate and land use scenarios
Health	To understand the current and future distribution and intensity of climate-sensitive health determinants, and predict outcomes that affect human and animal health in mountain regions
Mountain economies	To assess the value of mountain ecosystem services and how that value is affected by different forms of management
Society and global change	To understand the environmental, economic, and demographic processes linking rural and urban areas in mountain regions, as well as those leading to urbanization, peri-urbanization and metropolization

environmental change although at present the principal ones are climate, land use change, and their interactions. In a mountain environment context, environmental change will affect the capacity of landscapes to continue to provide services not only for resident populations but also for dependent populations beyond the mountains' extent. Changes in temperature and precipitation regimes, and the frequency of extreme events, will have severe repercussions on the physical character and on the biological and human communities of mountain areas. Glaciers, snow cover, water storage and flow, are unique features of mountainous areas; however, any changes affecting them will in turn impact on lowland areas. The range of environmental conditions present in mountain regions has resulted in mountains playing a significant part in the conservation of global biodiversity. Environmental changes will affect not only the species present, but also ecosystem functions such as biochemical cycling and habitat provision. Climate changes, through the alteration of the frequency of extreme events, may pose a threat to life and property in mountain regions, create new health hazards for humans and their domestic animals, and cause severe impacts on mountain economies and livelihoods. Land use, one of the major drivers of change globally, is subject to external forces such as climate and global markets. The response to change will not only require increased understanding of how drivers operate in mountain environments but also the appropriate institutional capacities to react.

Although many textbooks and specialized books have been written about mountain environments (e.g. Gerrard, 1990; Barry, 1992; Funnell and Parish, 2001; Parish, 2002) there has been only one so far for the Mediterranean mountains (McNeill, 1992). This first chapter is an introduction to the mountain environments of the Mediterranean Basin. It will set the scene and place the mountains of the basin in the global context while at the same time providing a brief overview of the various subjects illustrated in this book, and explain the organization of its content.

1.2 Setting the scene

Unquestionably mountains constitute the backbone of the whole Mediterranean region, including the largest islands in the basin (Figure 1.1). McNeill (1992, p. 1) in the only book dedicated to the Mediterranean mountains to be published so far, wrote: 'The beauty of the Mediterranean mountains is in way a sad one. Skeletal mountains and shell villages dot the upland areas of the Mediterranean world, dominating the physical and social landscape.' Imposing massifs run from North to South such as the Pindos in Greece, the Apennines in Italy, and the Dinarids in Balkans, West to East such as the Atlas Mountains extending over 3500 km from North Morocco to Tunisia, and dominate the landscapes of Sicily, Sardinia, Corsica, Cyprus and Crete.

Definitions about what constitutes a mountain vary and include topography, climate, vegetation, constraints on agriculture, or length of growing seasons (Gerrard, 1990; Kapos, 2000). A working definition is the one by Price (1981): 'An elevated landform of high local relief, e.g. 300 m (1000 ft), with much of its surface in steep

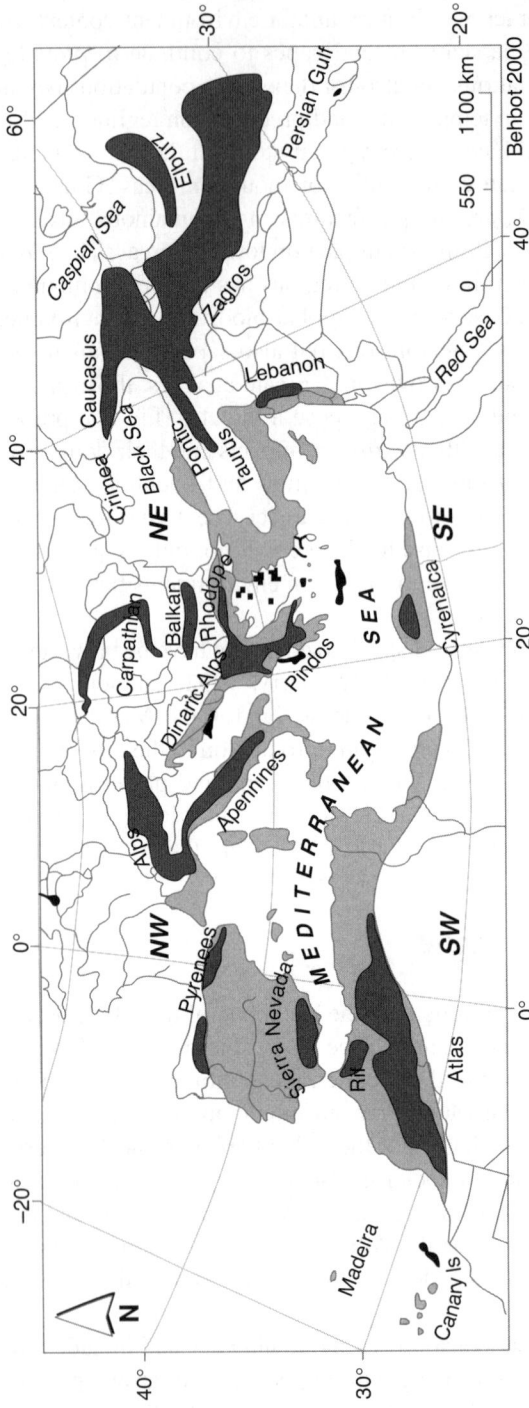

Figure 1.1 Delimitation of the Mediterranean biogeographical area with the major Mediterranean mountains. Reproduced from Blondel *et al.* (2010), with permission

Table 1.2 Percentage of mountainous land within Mediterranean countries (Regato and Salman, 2008)

Mediterranean countries	Percent of mountainous area within a country				
	0–10	10–25	25–50	50–75	75–100
Andorra, Bosnia and Herzegovina, Italy, Lebanon, Macedonia, Montenegro					▓
Albania, Greece, Morocco, Serbia, Slovenia, Turkey, Palestinian Territories				▓	
Croatia, Cyprus, Israel, Portugal, Spain			▓		
Algeria, France, Jordan, Syria, Tunisia		▓			
Egypt, Libya, Malta	▓				

slopes, usually displaying distinct variations in climate and associated biological phenomena from its base to its summit.' According to this definition an attempt to answer the question how large is the mountain area globally came up with the figure of 24.3% of total land area. The figure excludes the major plateau areas, and all land area outside Antarctica (Kapos *et al.*, 2000). Mediterranean mountains cover some 1.7 million km^2, equivalent to 21% of the combined surface area of all the countries concerned, and are home to 66 million people, representing 16% of the region's total population. Mountains occupy more than 50% of land in many Mediterranean countries (Table 1.2), seven of which are among the top 20 mountainous countries in the world. Morocco is an example of such country with a high percentage of mountainous land, where four major mountain ranges – the Rif, the Middle Atlas, the High Atlas and the Anti-Atlas – occupy 15% of its territory (Radford *et al.*, 2011)

The delineation of the Mediterranean mountains is a more complex issue since it is directly related to the delineation of the Mediterranean area itself, which has been the topic of debate for decades. For the latter the criteria used have been floristic, climatic or bioclimatic (see Blondel *et al.*, 2010). In a recent attempt at an environmental classification for the whole of Europe (Metzger *et al.*, 2005), Mediterranean mountains were recognized as separate entities influenced by both the Mediterranean zone they are situated in, and a distinct mountain climate. The class 'Mediterranean mountains' encompasses low and medium mountains in the northern part of the basin and high mountains in the southern part. As Blondel *et al.* (2010) state, 'there is no satisfactory answer to the question of what is a "Mediterranean mountain", as compared to a mountain range simply marking a regional boundary.' Taking into account the various delineations and classification schemes applied in the Mediterranean Basin, we broadly distinguish three mountain categories in this book from the heart to the periphery of the basin:

- Mountains at the very heart of the Mediterranean either due to their geographical position or their moderate altitude, including the Sierra Nevada, Cyrenaica, mountains of the Peloponnese and those on the islands of Sicily, Corsica, Crete and Cyprus.

- Mountains that are considered outside the influence of, or on the periphery of the Mediterranean, for example the Alps, Mercantour, the North Dinarids, and the North of Anatolia (Ozenda, 1975).

- Mountains included in the various delineations of the Mediterranean area that have biotic affinities to the basin, such as the mountains of the Canary islands.

This book focuses on the first category, since extending this volume to cover all the massifs associated with the Mediterranean Basin would be a huge task. However, there are limited references to the other two categories (see Chapters 3 and 4).

Mediterranean mountains exhibit many similarities in their biotic, ecological, physical and environmental characteristics but also significant differences. They have always been inextricably linked to their surroundings, providing, for example, cities and coastal areas with invaluable resources including water, timber and even labour (Benoit and Comeau, 2005). Relief in the Mediterranean is affected more by erosional processes than by glacial abrasion, compared to other European mountains; Mediterranean mountains receive more precipitation compared to the surrounding lowlands and are sources of various important rivers. In general, mountains in the northern part of the Mediterranean are lower than those in the southern Mediterranean. Primary and secondary shrub formations (e.g. maquis, garriga, carrascal, phrygana, shibliak) are very common, with distinct differences between north and south. Floristic composition, and the level and concentration of endemism vary (see Chapter 6). Other differences include human colonization patterns, historic land uses and current anthropogenic pressures. For example, tourist impact is greater in northern than in southern Mediterranean mountains, whereas grazing follows the opposite trend. This book addresses these characteristics and examines the major environmental changes that the mountains experienced during the Quaternary period.

1.3   The character of the Mediterranean mountains

The separation of the African and European plates around 150 million years ago resulted in the formation of the Mediterranean Basin. Throughout the last part of the last ice age, c. 20 000 years ago, the climate of the area was significantly drier and cooler than it is today. The mountains of the Mediterranean have changed significantly since the end of the last ice age c. 12 000 years ago, due to sea-level rise, and in turn the biogeographical characteristics of the mountains have been altered. In *Chapter 2*, faunal and palynological evidence is examined to provide a picture of the changes in biota and the environment through the Quaternary, with emphasis on changes, including human impacts, during the last 12 000 years.

Many of the Mediterranean mountains supported glaciers during the Pleistocene, and some glaciers and ice patches still survive today, as discussed in *Chapter 3*. Cirques, U-shaped valleys, arêtes, roches moutonées, glacial lakes and moraines present today in the mountains of the region are all evidence of these glaciations that have shaped the landscape. Today the majority of the glaciers present are restricted to the highest mountains such as the Pyrenees, the Maritime Alps and the mountains of Turkey. However, several glaciers also exist in lower mountain areas, such as central Italy and in Montenegro and Albania (see Chapter 3).

The presence of a ring of mountains around the Mediterranean Basin, with the exception of the southeastern part, is a result of the collision between the African and the Eurasian plates. Some of the mountains, like the Sierra Nevada, are underlain by Archaean structures while others, for example the Maritime Alps, by more recent Miocene deposits. The Atlas, Rif, Baetic Cordillera, Cantabrian Mountains, Pyrenees, Alps, Apennines, Dinaric Alps, Hellenides, and Balkan and Taurus mountains are all products of the alpine orogeny, while the mountains of Portugal and western Spain, as well as Sardinia, are of Hercynian origin. Many of these such as the Sierra Nevada, Etna, Taurus, High Atlas and Mount Lebanon reach over 3000 m while there are several active volcanoes in the area. *Chapter 4* provides an overview of tectonic setting and landscape development of the Mediterranean mountains, including the range of geological substrates encountered. Mountains are high-energy environments characterized by great instability and variability. This is well demonstrated in the soils of Mediterranean mountains, where a variety of raw lithomorphic soils, particularly in the high alpine zones, is present.

The altitudinal and continental position as well as latitude and topography of mountains exert an influence on climate (Barry, 1992). Mountain areas worldwide contain the sources of all major rivers, and those of the Mediterranean are no exception. Rivers like the Guadalquivir, Ebro, Rhône and Po have their source/origins on some of the most important mountains in the area, contributing water to dry lowlands. However, climate and hydrology in the Mediterranean mountains receive little attention compared to other mountain massifs worldwide. *Chapter 5* provides a comprehensive overview of interaction of the major hydrological and meteorological processes in Mediterranean mountain areas. This overview includes snowmelt, run-off and floods, water fluxes and water balance, hydrometeorological coupling and modelling. It reviews recent research in the field and illustrates key interactions from a range of mountainous regions in the Mediterranean. Emphasis is given to human impacts, assessment of mountain water resources, conservation and water quality.

Due to their altitudinal range, mountains contain a wide range of environments in a short distance and therefore support more habitats than equivalent lowland areas. Although in absolute terms the number of species is smaller compared to lower areas, Mediterranean mountains support floras of special interest. They all contain a high number of endemic species, with southern mountains having a higher percentage of endemism than those in the north. Some endemics are relictual, whereas others are more recent as a result of specific and localized factors such as discrete orogenies and rare substrates. *Chapter 6* discusses the mountains' biogeographical affinities and peculiarities in terms of biota and habitats. Emphasis is given to the distinct altitudinal zonation and the main vegetation types encountered in the

Mediterranean mountains as well as the adaptations of biota to high-altitude environmental conditions. In addition the pressures on mountain biodiversity and the efforts at national and international level for its protection are discussed.

Throughout the Mediterranean Basin, human identities and cultures in mountain areas are diverse both historically and currently. Mountains provide refuge not only to plant and animal species but also to people trying to escape from invaders, and they form a distinctive component of many peoples' cultural identity. In the interior mountains of each Mediterranean country, human populations are still extraordinarily distinct linguistically and behaviourally (McNeill, 1992). People have always had spiritual connections with and cultural roots in the most imposing massifs of the Mediterranean, while traditional management systems have resulted in cultural landscapes with unique biodiversity, cultural and socioeconomic values. The link between culture and Mediterranean mountain environments is dealt with in *Chapter 7*.

Environmental and economic change is no stranger to the Mediterranean mountain environments (for the role of natural forces see Chapters 3, 4 and 5). In their quest for timber, fuel and minerals, humans have left their irreversible mark on these mountains. The increasing reliance of human communities on mountains for various services has led to a magnitude and rate of change that threatens to overwhelm mountain ecosystems and hence the communities they support. Currently Mediterranean mountain environments, and the human communities that live and work there, face unprecedented threats from social, economic and environmental forces of change. These same forces also bring exciting opportunities for the integration of knowledge and expertise to achieve sustainable solutions for future development of these areas. The first part of *Chapter 8* provides a historical analysis of land uses (grazing, transhumance, terrace cultivations), the drivers (economic, social and environmental) that determine the pattern of change, and how those vary across the Mediterranean mountains. The second part provides a description of the current situation in some of the most mountainous countries in the Mediterranean and discusses the impacts of land use change and future challenges.

Climate change poses a number of potential risks to ecosystems globally. Although the impacts cannot as yet be predicted with certainty, mountain systems are particularly sensitive to changes in climate, supported by past evidence (geological era) of vegetation zone shifts (Beniston, 2000). Mediterranean mountains are under threat from climate change, which affects directly or indirectly different key features, such as biodiversity, snow cover, glaciers, run-off processes and water availability. *Chapter 9* reviews recent and future trends in Mediterranean mountain climate. It provides an assessment of temperature, precipitation, and spring precipitation changes in Mediterranean mountains under different emission scenarios and Atmosphere-Ocean-Coupled General Circulation Models. The implications of predicted climate change for both human and physical features and synergies with other pressures on mountain environments are discussed, focusing on the cryosphere, hydrosphere and biodiversity.

The Mediterranean has been exploited for thousands of years, and some scholars claim that it has proven to be resilient (Grove and Rackham, 2001). Mediterranean mountain landscapes in the past achieved equilibrium among the principal activities (agriculture, pastoralism and forestry), which also promoted environmental protection (see Chapter 8). This is no longer the case, bringing severe consequences for mountain resources. Do the socioeconomic changes of the last 50 years threaten resources and cultural identities? Have we exceeded carrying capacity and can we reconcile development and conservation? What are the challenges and the options for mountain environments and the communities that depend on them. The last chapter of this book (*Chapter 10*) underlines the differences and similarities of Mediterranean mountains. It provides a synthesis of the challenges that the mountains are facing, as outlined in this chapter and addressed in subsequent relevant chapters, and places those in the light of potential future development. This concluding chapter emphasizes the need to ensure the ecological health and the economic and social improvement of mountain areas through an ecosystem-based approach, which will take into account both the physical environment and the livelihoods of mountain and lowland communities alike.

References

References marked as bold are key references.

Barry, R.G. (1992) *Mountain Weather and Climate*. London: Routledge.

Beniston, M. (2000) *Environmental Change in Mountains and Uplands*. Arnold.

Benoit, G. and Comeau, A. (2005) *A Sustainable Future for the Mediterranean: The Blue Plan's Environment and Development Outlook*. Earthscan.

Björnsen Gurung, A. (ed.) (2005) *GLOCHAMORE: Global Change and Mountain Regions: Research Strategy*. Report of the EU Framework Program 6 (Contract No. 506679): Global Change and Mountain Regions: An Integrated Assessment of Causes and Consequences (November 2003 – October 2005).

Blondel, J., Aronson, J., Bodiou, J.-Y. and Boeuf, G. (2010) *The Mediterranean Region: Biological Diversity in Space and Time*. Oxford University Press.

Funnell, D.C. and Parish, R. (2001) *Mountain Environments and Communities*. London: Routledge.

Gerrard, A.J. (1990) *Mountain Environments: An Examination of the Physical Geography of Mountains*. MIT Press.

Grove, A.T. and Rackham, O. (2001) *The Nature of the Mediterranean Europe. An Ecological History*. New Haven, CT: Yale University Press.

Kapos, V., Rhind, J., Edwards, M., Price, M.F. and Ravilious, C. (2000) Developing a map of the world's mountain forests. In: Price, M.F. and Butt, N. (eds), *Forests in Sustainable Mountain Development: A State-of-Knowledge Report for 2000*. Wallingford, UK: IUFRO Publishing/CABI International, pp. 4–9.

Körner, C. and Ohsawa, M. (coords., lead authors) (2005) Mountain systems. In: Hassan, R., Scholes, R. and Ash, N. (eds), Millennium Ecosystem Assessment. Island Press, Chapter 24.

McNeill, J.R. (1992) *The Mountains of the Mediterranean World.* **Cambridge: Cambridge University Press.**

MEA (2005) *Millennium Ecosystem Assessment.* Hassan, R., Scholes, R. and Ash, N. (eds). Washington, DC: Island Press.

Metzger, M.J., Bunce, R.G.H., Jongman, R.H.G., Mücher, C.A. and Watkins, J.W. (2005) A climatic stratification of the environment of Europe. *Global Ecology and Biogeography* 14: 549–563.

Ozenda, P. (1975) Les limites de la vegetation mediterranéenne en montagne, en relation avec le projet de Flora Mediterranea. In: Guinochet, M., Guittonneau, G., Ozenda, P., Quezel, P. and Sauvage, Ch. (eds) La flore du bassin mediterranéen essai de systematique synthetique. Paris: CNRS, pp. 355–343.

Parish, R. (2002) *Mountain Environments.* Routledge.

Price, L.W. (1981) *Mountains and Man: a Study of Process and Environment.* University of California Press.

Price, M.F. (ed.) (2004) *Conservation and Sustainable Development in Mountain Areas.* Gland, Switzerland and Cambridge, UK: IUCN.

Price, M.F. (ed.) (2007) *Mountain Area Research and Management: Integrated Approaches.* Earthscan, London.

Radford, E.A., Catullo, G. and de Montmollin, B. (2011) *Important Plant Areas of the South and East Mediterranean Region: Priority Sites for Conservation.* IUCN, Plantlife, WWF.

Regato, P. and Salman, R. (2008) *Mediterranean Mountains in a Changing World; Guidelines for Developing Action Plans.* **IUCN.**

2

Quaternary environmental history

A.M. Mannion

2.1 Introduction

The uplands bordering the Mediterranean Sea (see Chapter 1, Figure 1.1) were formed when sediments that accumulated in the ancient Tethys Ocean were subject to heat and pressure as the North African and Eurasian plates collided. This occurred between 100 million and 60 million years ago leaving a shrunken Tethys, namely the Mediterranean Sea, as configured today. Thereafter these uplands were subject to plant and animal colonization including the newly evolved mammals and angiosperms, soil formation, and denudation processes in the Tertiary. The warm tropical/subtropical climate of the Tertiary gradually deteriorated, culminating with the onset of glaciation in northern hemisphere, the first cold stage of the Quaternary, now redefined by the International Commission on Stratigraphy as commencing c. 2.6 million years ago (Gibbard *et al.*, 2009). This climatically diverse period was characterized by repeated cycles comprising cold and warm stages that, at their extreme expression, were glacials and interglacials. In general during the early part of the Quaternary, before 900 000 years ago, cold/glacial stages were roughly of the same duration as warm/interglacial stages, but in the later part of the Quaternary the cold/glacial stages were considerably longer than warm/interglacial stages and all were characterized by internal climatic and ecosystem dynamism. Thus the last 2.6 million years has been a period of considerable environmental change on relatively short geological timescales. This was a global phenomenon, with evidence primarily derived from continental glacial and cave deposits, exposed and offshore marine sediments and a few deep lake sediment sequences.

How Quaternary environmental change progressed in the Mediterranean mountains is difficult to determine, not least because evidence of early cold/glacial stages and warm/interglacial stages has been destroyed by the glaciers of the last ice advance. Moreover, some regions have not been well investigated, for example the Balkans, which creates spatial gaps in current knowledge. For most of the

Mediterranean Mountain Environments, First Edition. Edited by Ioannis N. Vogiatzakis.
© 2012 John Wiley & Sons, Ltd. Published 2012 by John Wiley & Sons, Ltd.

Quaternary period it is essential to refer to sedimentary records of outcrops and cores from the Mediterranean region and to the long sedimentary sequences of lakes in Greece, Albania/Macedonia, Italy, Israel and southeast France. There is much more local evidence for environmental change during the last glacial period and the Holocene, that is, the last 100 000 years in the Mediterranean mountains. Much of this comprises glacial deposits and landforms, which are examined by Hughes (see Chapter 3) and to which only brief reference will be made here. In addition there are numerous palaeoecological investigations of lake sediments and peats that relate to the last 12 000–10 000 years. Pollen analysis in particular has been widely used to reconstruct Holocene vegetation communities in these upland mainland regions and in the Mediterranean Islands (see Mannion, 2008; Rackham, 2008, for reviews of the latter).

2.2 The pre-Quaternary period

Towards the end of its 50×10^6 year existence the supercontinent of Pangea began to fracture, c. 200×10^6 years ago. The Atlantic Ocean came into existence as the Americas moved westwards, while an extensive waterbody, the Tethys Ocean, formed between the separating African and Eurasian plates. By c. 120×10^6 years ago those same plates began to converge, gradually reducing the extent of the Tethys so that only a remnant was present by 40×10^6 years ago. This was the Mediterranean Sea. By this time the converging plates had also caused the uplift of sediments around the edge of the basin to produce the Mediterranean mountains, discussed herein, as well as the Alps. Denudation processes operative on land produced sediment, which was deposited first in the Tethys Ocean and then in the Mediterranean Sea. As these sediments consolidated they encapsulated physical, chemical and biological components that reflected environmental conditions at the time of deposition. The particle size, mineral and organic contents of the sediments (e.g. black shales or sapropels) plus the remains of organisms such as foraminifera, ostracods, diatoms, radiolarians, coccolithophores, pollen grains and marine molluscs all contribute to understanding past environmental change; chemical signatures in the fossils, especially foraminifera, include oxygen isotopes, which have facilitated the construction of stratigraphies for correlation between cores and for sea surface temperature reconstruction (see Mannion, 1999; Anderson *et al.,* 2007, for details about theory, method and applications). Thus marine sediment cores plus a number of uplifted sediments that are now dry land provide evidence on which to examine late Tertiary and Quaternary environmental change in the region as a whole.

 Of particular significance is the Messinian Salinity Crisis (see CIESM, 2008, for a recent review of this contentious issue), which occurred c. 6×10^6 years ago. This involved closure of the link between the Atlantic Ocean and the Mediterranean Sea, that is, in the area of the Gibraltar arc (Warny *et al.,* 2003), either due to volcanic activity (Duggen *et al.,* 2003) or through tectonic uplift (Jolivet *et al.,* 2006) but not because of climatic change (Fauquette *et al.,* 2006). Erosion causing fluvial canyon formation and desiccation occurred in a dry basin which reached depths of 1.5 km and in which extensive evaporites were deposited. In a recent review of research on this contentious issue, Ryan (2009) states that it was 'an extraordinary event in which 5% of the dissolved salt of the oceans of the world was extracted in a fraction of a million years to form a deposit more than 1 million km^3 in volume ... The dolomite, gypsum, anhydrite and halite in the drill cores paint a surprising picture of a Mediterranean desert lying more than 1 km below the Atlantic Ocean with brine pools that shrank and expanded by the evaporative power of the sun.' Much of the salt bed formation was influenced by orbital forcing, that is insolation changes related to the path taken by the Earth as it revolves around the Sun and the way in which the Earth revolves on its axis, just as the advance and retreat of the cold/glacial and warm/interglacial stages were influenced during the Quaternary period (see below).

Some of these marginal Messinian evaporites deposited during the first step of the crisis (Clauzon et al., 1996) were later uplifted by tectonic activity to form dry land as in Messina, Sicily, northeast Libya, Italy and southern Spain. The time frame for this process is contentious, with estimates varying between 1.5×10^6 years (Butler et al., 1999) and a relatively sudden event over 0.26×10^6 years (Clauzon et al., 1996; Krijgsman et al., 1999). Such conditions prevailed for more than 0.5×10^6 years, during which time there were several subphases of erosion. These numbered between three and five (Gargani and Rigollet, 2007). The subsequent high sea-level of the Atlantic Ocean allowed saltwater with a mixed flora and fauna to re-enter the Mediterranean and may have been linked with stream piracy (Blanc, 2002) which helped breach the divide. According to Favre et al. (2007), vegetation maps based on pollen assemblages from 30 locations throughout the Mediterranean Basin indicate that open habitats with steppe communities prevailed to the south, except some savanna in the Nile region, while the north of the basin was characterized by forest mosaics that varied in composition with relief.

Although this information reflects environmental change in the Mediterranean Basin itself, the corollary is that the mountain regions within the basin were also experiencing environmental change but there is much less palaeoenvironmental information from the mountain regions themselves. Willett et al. (2006) report that the Messinian event was paralleled by increased erosion in the Alps, which they attribute to increased rainfall. This was probably the case in the Mediterranean mountains discussed in this book, though not all. For example, Babault et al. (2006) argue that the Ebro River basin of the Pyrenees became connected to the Mediterranean Basin after rather than during the Messinian event, on the basis that there are no incised 'canyons' typical of Messinian age landforms.

2.2.1 The Pliocene 5.3 to 2.6×10^6 years ago

Relative climatic stability that characterized the Tertiary period globally was coming to an end as a cooling trend became established in a still warm and humid world.

Evidence for regional climatic change that would have affected Mediterranean mountains is indirect, deriving from marine sediment cores from the Mediterranean Sea. Based on pollen assemblages, Suc (1984), Suc *et al.* (1995) and Popescu *et al.* (2010) have confirmed the tripartite division of the Pliocene suggested by Zagwijn (1960), though the third subdivision is now designated as the Quaternary, as discussed above.

5.2 to 3.37 $\times 10^6$ years ago

The evidence for this time period derives from a variety of sites and as might be expected it reveals considerable variation throughout the Mediterranean region. Suc *et al.* (1995) describe six spatial units of which those numbered 2, 3 and 4 below are particularly relevant to mountain environments:

1. In the Atlantic region warm, humid conditions prevailed and subtropical and warm-temperate tree species dominated with ericaceous shrubs.

2. South of latitude 42°N ecosystems were dominated by xerophytic species including species characteristic of desert margins, which suited the seasonally dry climates.

3. In the northwest, i.e. the Pyrenees to Central Italy including the Apennines, three forest belts developed. Spruce (*Picea*) and fir (*Abies*) dominated high altitudes; middle altitudes were characterized by cedar (*Cedrus*) with *Sequoia*-type species and with *Cathaya* dominant at low latitudes. Such assemblages indicate an altitudinal gradient towards cooler and humid conditions. In this region, lowlands were characterized by a Mediterranean-like vegetation mosaic.

4. In the northeast, temperate forest species alternated with grassland species.

5. Savanna with a few desert species and sparse riparian tropical gallery-forests occupied the Nile delta area.

6. Wetlands in Portugal, Catalonia and Romania were dominated by *Taxodium*-type, *Nyssa, Myrica* and *Cyrillaceae-Clethraceae.*

Recent work by Popescu *et al.* (2006, 2010) provides further evidence for early Pliocene environmental change. Pollen analysis of a core from the Lupoaia section, a delta system of lignites near the Carpathians in southwest Romania, shows that thermophilous elements and trees reacted to orbital forcing parameters. In particular the former increased during eccentricity minima while the tree line descended altitudinally during maxima when overall temperatures were lower. Precession also had an impact in relation to humidity, which encouraged the spread of cypress swamps and marshes during precession maxima. Although the Carpathians are beyond the scope of this book these results, plus similar findings from Sicily (e.g. Hilgen, 1991), reflect the importance of orbital characteristics as influences on regional/global climates, which in turn affect vegetation communities and earth-surface processes. Undoubtedly the Atlas, Apennines, Albanian Alps, etc. were similarly influenced by the climatic changes of the early Pliocene.

3.37 to 2.62 × 10⁶ years ago

Conditions described above continued during the latter part of the Pliocene. Declines in some thermophilous species indicate declining temperatures, increased annual temperature gradients and more pronounced seasonality. Toward the close of this period there is evidence for glaciation. This is reflected in Figure 2.1, which shows the oxygen-isotope stratigraphy for the last 3.2 million years. Between 3.37 and 2.62 million years ago warm-cool oscillations occurred, with an overall trend for cooling, but c. 2.62 million years ago the amplitude of the cycles deepened. This date correlates with the onset of glacial conditions globally (see below).

2.3   The Quaternary period

The redefined base of the Quaternary period encompasses what was the final stage of the Pliocene; it is the third of the stages identified by Suc (1984) and Suc *et al.* (1995) and coincides with the onset of glaciation on a global scale. The type site is situated in the Mediterranean in the sedimentary sequence at Monte San Nicola, Sicily, and is the base of the former Pliocene unit known as the Gelasian. The stratigraphy of this sequence is given in Figure 2.2, which shows that considerable depositional changes occurred. The former Quaternary type site at Vrica, Calabria, the base of which is dated at 2.47 million years old, is now considered to encompass the second stage of the revised Pleistocene, which is itself a combination of the Quaternary and the Holocene (Gibbard *et al.*, 2009). The Quaternary is characterized by recurring climatic cycles comprising cold and warm stages of almost equal duration at the beginning, and then long cold stages and short warm stages, the temperature difference between which intensified c. 900 000 years ago as shown in the oxygen-isotope stratigraphy of Figure 2.1. In relation to Mediterranean mountain environments these climatic cycles have caused considerable ecological and environmental change as tree lines have marched up and down mountains, and ecosystems at all altitudes and latitudes have reconfigured in terms of species composition.

2.62 to 1.8 × 10⁶ years ago

Fossil pollen assemblages in marine cores from the northern Mediterranean presented by Suc (1984), Suc *et al.* (1995) and Popescu *et al.* (2010) show that the

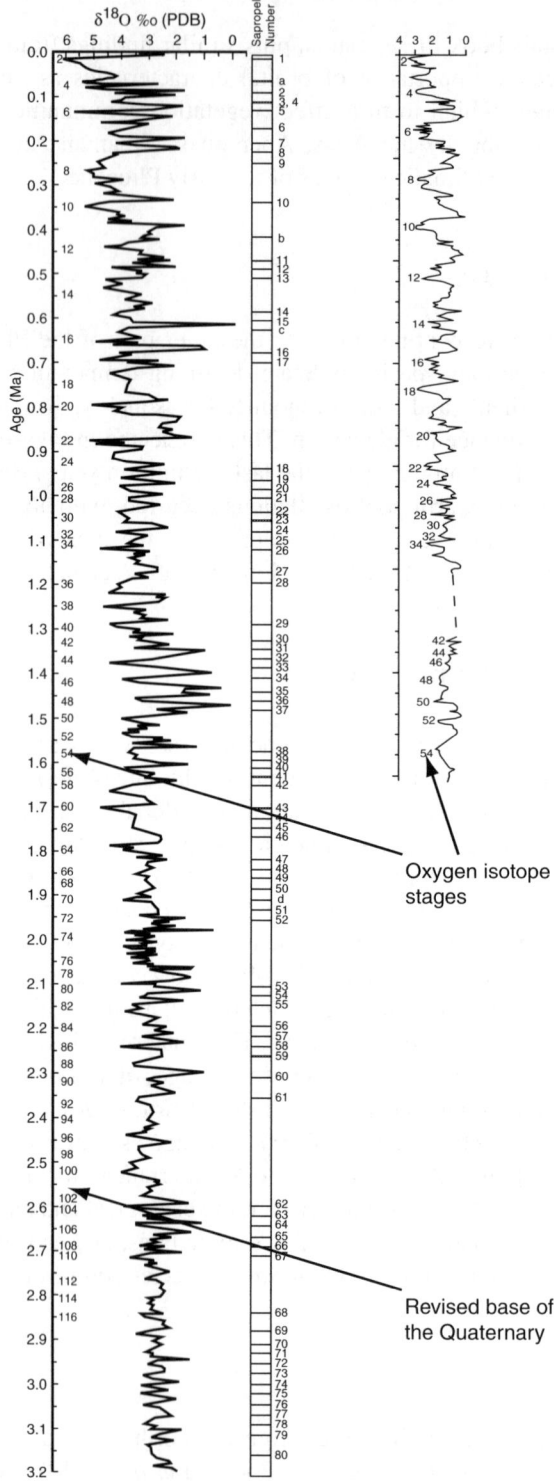

Figure 2.1 Oxygen-isotope stratigraphy of Ocean Drilling Programme cores 967 (Eastern Mediterranean) and 976 (Western Mediterranean). Reproduced from Kroon *et al.* (1998) and Nebout *et al.* (1999), with permission

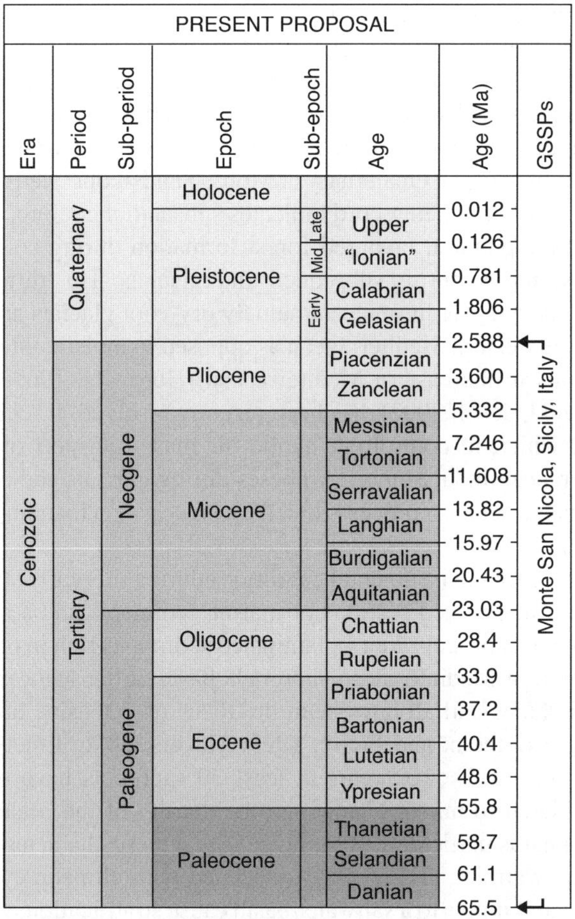

Era	Period	Sub-period	Epoch	Sub-epoch	Age	Age (Ma)	GSSPs
Cenozoic	Quaternary		Holocene			0.012	Monte San Nicola, Sicily, Italy
			Pleistocene	Late	Upper	0.126	
				Mid	"Ionian"	0.781	
				Early	Calabrian	1.806	
					Gelasian	2.588	
	Tertiary	Neogene	Pliocene		Piacenzian	3.600	
					Zanclean	5.332	
			Miocene		Messinian	7.246	
					Tortonian	11.608	
					Serravalian	13.82	
					Langhian	15.97	
					Burdigalian	20.43	
					Aquitanian	23.03	
		Paleogene	Oligocene		Chattian	28.4	
					Rupelian	33.9	
			Eocene		Priabonian	37.2	
					Bartonian	40.4	
					Lutetian	48.6	
					Ypresian	55.8	
			Paleocene		Thanetian	58.7	
					Selandian	61.1	
					Danian	65.5	

(The table is headed by the banner: **PRESENT PROPOSAL**)

Figure 2.2 The revised base of the Quaternary period, a Global Stratotype Section and Point (GSSP), as accepted in 2009 by the International Union of Geological Sciences (IUGS). Reproduced from Gibbard *et al.* (2009), with permission

onset of cold/glacial–warm/interglacial cycles was characterized in the Mediterranean lowlands by steppe vegetation dominated by *Artemisia* and *Ephedra* during the former while forests prevailed during the shorter warm periods. Similarly, in the eastern Mediterranean, pollen assemblages from the Hula Basin, a drained/infilled lake basin c. 70 m asl (above sea level) in a rift valley that contains sediments derived from two mountain regions, the Golan Heights and Naftali Mountains, indicate intense cooling between 2.4 and 2.6×10^6 years ago (Horowitz, 1989). However, in the southern Mediterranean Basin the pollen assemblages indicate that environmental change was less pronounced, suggesting that temperature depression was of insufficient magnitude to prompt major vegetation change. As Hughes (Chapter 3) points out there is only limited evidence for glaciation during the early

Quaternary in the Atlas Mountains and there is no dating control. Nevertheless it is important to note that Suc (1984), Suc et al. (1995), Suc and Popescu (2005) and Popescu et al. (2010) date the onset of cold/glacial stages to 2.6×10^6 years ago, which is in agreement with dating of ice-derived material in Atlantic sediment cores (e.g. Shackleton et al., 1984; Haug et al., 2005) and is in accord with the dating of the newly agreed base of the Quaternary at Monte San Nicola, Sicily (see above). In contrast the Hula Basin pollen record indicates substantial cooling, which, coupled with increased precipitation, favoured forest formation during cold/glacial stages and steppe communities during subsequent interglacials. This difference between the western and eastern Mediterranean, namely dry-cool glacials and humid-warm interglacials in the western Mediterranean as opposed to moist-cool glacial and dry-warm interglacials in the eastern Mediterranean (Horowitz, 1969; Subally et al., 1999; Subally and Quézel, 2002), has been recently resolved. It relates to the monsoon influence, which in controlling humid air mass transport over the central-eastern Mediterranean induces moister phases during glacials and conversely drier phases during interglacials, a process forced by precession (Joannin et al., 2007a,b; Suc et al., 2010).

Further evidence for environmental conditions during this early stage of the Quaternary derives from exposed sections of marine sediments and a range of marine cores which contain sapropels, that is, laminated sediments rich in organic material. These central-eastern Mediterranean sapropels have earlier Miocene counterparts known as black shales and all have been the focus of extensive investigation (for commentaries see Béthoux and Pierre, 1999; Meyers, 2006; Emeis and Weissert, 2009). During the last 3×10^6 years at least 80 sapropels have been deposited, which are considered to mirror changes in the density of the ocean water and its ability to mix. What is particularly significant is that these characteristics are linked with the climatic warming and cooling associated with climatic cycles. Major increases in freshwater inputs to a saltwater basin cause stratification, which precludes mixing. Consequently surface layers remain oxygenated but deep layers become anoxic (i.e. depleted of oxygen). This kills benthic organisms and inhibits their decomposition; their remains become incarcerated in the accumulating sediment and thus enhance the organic content to produce a sapropel. Periods of increased rainfall would have generated an increased influx of freshwater (Hilgen, 1991; Suc et al., 2010) as would a greater volume of glacial meltwater deriving from the Alps or the Mediterranean mountains, or from sources bordering the Atlantic Ocean whose impacts would have fed into the Mediterranean via the Straits of Gibraltar. These processes would have operated during warm stages and would thus have promoted substantial sapropel formation, which was limited during cold stages. As Hilgen (1991) and Suc et al. (2010) suggest, based on evidence from sedimentary sequences from southern Italy (notably the Crotone sequence), the palaeoenvironmental record may reflect the response of vegetation to the strong interaction between precession and obliquity during climatic cycles. Such relationships are difficult to unravel but emphasize the complex links between global climate and the carbon cycle (Mannion, 2006).

What does this information mean in relation to the early (or later) Quaternary history of Mediterranean mountains? The mountains themselves may have been divested of evidence for such change due to the tremendous erosional power of a succession of ice caps and glaciers but the climate reconstructions facilitated by ocean-sediment data reveal that the lands around the Mediterranean Basin, including the mountains, were susceptible to the global climatic cycles of the early (and later) Quaternary ups and downs of ice sheets and glaciers. Just as glaciers/ice caps waxed and waned, forest, steppe and even desert vegetation communities shifted latitudinally and altitudinally and changed in relation to species composition in the mountains. Such shifts may not have been uniform throughout the Mediterranean, as reported by Meyers and Arnaboldi (2008), whose work on isotope signatures in marine sediments suggests that wetter conditions, that is, warm stages, were more strongly expressed in the eastern Mediterranean than in the western Mediterranean. This trend may also indicate increased continentality during cold/ice stages in the eastern Mediterranean mountains, that is, increased extension of glaciers.

1.8 to c. 1×10^6 years ago

The base of this Middle Quaternary stage is defined on the basis of the stratigraphy at Vrica, Calabria, southern Italy. The sediments were deposited in deepwater and include a claystone overlying a sapropel deposit; the junction is dated at 1.81×10^6 years ago and is the base of the Calabrian (Van Couvering, 2004; Gibbard *et al.*, 2009). Marine sediment cores from the Mediterranean Basin provide the most comprehensive record of environmental change for this period because the record is continuous and not fragmented as is the case for that from terrestrial archives. In particular the oxygen-isotope stratigraphy and the fossil record provide important insights into the patterns and characteristics of the many climatic cycles that dominated this period. As Figure 2.1 shows, there are some 38 oxygen-isotope stages during this period, which represent some 19 climatic cycles in c. 800 000 years. Although the amplitude of many of these cycles was not sufficient to create glacial conditions, repeated warming and cooling would have affected the land cover and ecosystem dynamics throughout the Mediterranean Basin including mountain regions.

In terms of detail, pollen analytical evidence from Montalbano Jonico, in the Lucania Basin on the northeastern edge of the Southern Apennines of Italy, reported by Joannin *et al.* (2008), shows that during the period 1.25 to 0.90 million years ago (oxygen-isotope stages 37 to 23) significant changes in plant community composition occurred that were driven by climatic change, including variations in precipitation as well as temperature, and changing sea-level. Such changes parallel the oxygen-isotope stratigraphy recorded from the deep-water foraminifera *Globigerinoides ruber* and *Globigerina bulloides*, which are accepted indicators of sea temperature, as recorded in core from the ODP site 967 shown in Figure 2.1. Joannin *et al.* state that 'On the whole, the mesothermic vs. steppe [pollen] ratio and oxygen isotopic compositions made on *Globigerina bulloides* are correlated . . . . Hence, high percentages of mesothermic elements correlate with low values of $\delta^{18}O$, which indicate warm-temperate phases . . . In contrast, steppe elements together with high $\delta^{18}O$ values indicate colder phases'. In the Montalbino Jonico section 13 warm-temperate phases have been recognized and are dominated by *Quercus* spp.; the intervening steppe phases are dominated by *Artemisia*. Eight of these warm-temperate phases are correlated with interglacials, that is MIS (marine isotope stages) 23, 25, 27, 29, 31, 33, 35 and 37. Although any single study is limited in relation to the area it represents such studies provide insight into a period for which there is little land-based evidence. Overall, it is generally considered that these and other cyclical environmental changes recorded in Mediterranean (and other ocean sediment) cores are due to the orbital forcing parameters of obliquity and precession, and that the latter is a key component in precipitation receipts, which is also a major factor in sapropel formation (see above).

   Further information on this middle stage of the Quaternary derives from the Hula Basin, Israel. Pollen analysis (Horowitz, 1989) of sediments from this infilled lake basin shows that climatic cycles were characterized by forest during wetter and cooler glacial stages, and as conditions became drier and warmer maquis communities developed and eventually steppe during interglacials (see comments above re moisture availability being reduced in the eastern Mediterranean region when compared with the western Mediterranean region due to monsoon effects). As this middle stage of the Quaternary drew to a close the last remaining elements of Tertiary tropical communities disappeared, at least in the central part of the Mediterranean as is indicated by the Montalbano Jonico (southern Italy) pollen data discussed above. In a recent review of literature from the Iberian peninsula Mijarra *et al.* (2009) suggest that the last tropical taxa in the eastern basin also disappeared probably due to the intensification of glacial cyclicity, which involved lower temperatures during the cold stages. The implication for Mediterranean mountains, especially in the north and west, is that temperature-sensitive species disappeared from high elevations as treelines receded during cold stages. Mountain vegetation communities in the south were less acutely affected but were unlikely to have remained unaltered. As for the east, relatively few records and the vexed issue of monsoonal influence (see above) make any general conclusions difficult.

1×10^6 to 20×10^3 years ago

Evidence for environmental change during the later part of the Quaternary period is more widespread than that for earlier times. In particular there are several deep sedimentary sequences from lake basins in the Mediterranean region that have been investigated and that provide an additional and complementary perspective on this interval, besides marine sediments. The locations of these lake basins are given in Figure 2.3 and further detail is given in Table 2.1. There are no long records from the south of the Mediterranean Basin. Moreover, none of those from the north or east of the basin are at high altitude and so they do not directly record changes in

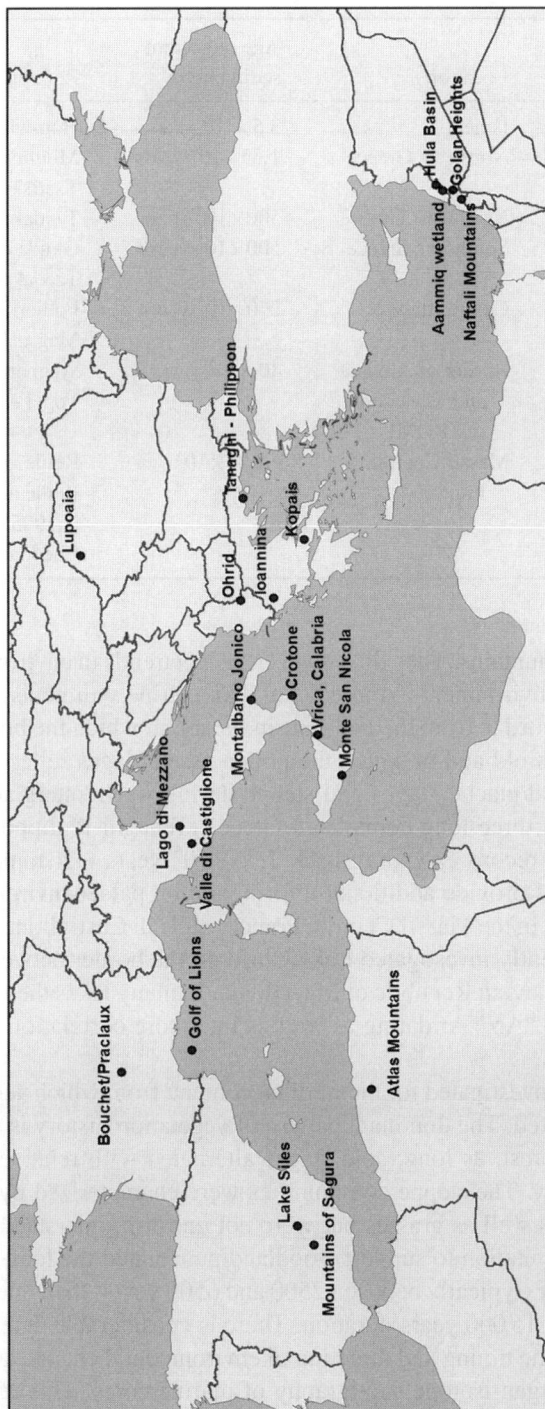

Figure 2.3 Map of sites mentioned in text

Table 2.1   Deep lake sediments in the Mediterranean Basin

Lake or basin	Location	Age of bottom sediments	References
Hula Basin	Israel	3.5×10^6 years	Horowitz, 1989
Tenaghi-Philippon	Northeast Greece	1.35×10^6 years	Mommersteeg *et al.,* 1995
			Tzedakis *et al.,* 2006
Ioannina	Northwest Greece	500×10^3 years	Tzedakis et al, 2003
Kopais	Southeast Greece	500×10^3 years	Okuda *et al.,* 2001
			Tzedakis, 1999
Valle di Castiglione	Central Italy	250×10^3 years	Follieri *et al.,* 1988
			Magri, 1989
Ohrid	Border of Albania and Macedonia (FYROM)	40×10^3 years	Wagner *et al.,* 2009
			Vogel *et al.,* 2010
Bouchet/Praclaux	Massif Central, France	c. 500×10^3 years	Reille *et al.,* 1988
			Reille and de Beaulieu, 1995
			Cheddadi *et al.,* 2005

Mediterranean mountains. They do, however, reflect trends in environmental change in the terrestrial environment in more detail than marine sediments.

The longest record is from the Hula Basin, Israel, in which the bottom sediments are 3.5×10^6 years old and in which the pollen assemblages reflect shifts between forests during cold/glacial stages and steppe during warm/interglacial stages (see above). There are three long cores from Greece: Tenaghi-Philippon – which provides the longest record extending back 1.35×10^6 years – Ioannina and Kopais. Three further sites provide additional information on palaeoenvironments: Lac du Bouchet/Praclaux in the Massif Central, France; Valle di Castiglione in central Italy; and the more recently investigated Lake Ohrid on the border between Albania and the Former Yugoslavian Republic of Macedonia. In many lakes there are tephra layers that facilitate $^{40}Ar/^{39}Ar$ dating as well as inter-core correlation and correlation with marine cores.

All have been investigated for their pollen content, from which vegetation history can be reconstructed. The dominant pattern of vegetation history is that of alternating steppe and forest, as long, cold stages alternated with relatively short, warm stages respectively. The steppe communities were characterized by *Artemisia and* Chenopodiacae as well as grasses but were not uniform; interstadials of sufficient warmth and precipitation to support woodland punctuated the long cold stages but were much shorter (typically between 2500 and 6500 years) than interglacials of between 10 000 and 15 000 years duration. There is considerable agreement between cores in terms of the timing and direction of environmental change, and between the cores and the oxygen-isotope stratigraphy of marine cores. This reflects the overall influence of climate within the region, not least climatic change as driven by

Milankovich orbital forcing characteristics. There is also evidence for environmental change during both cold and warm stages, indicating heterogeneity rather than homogeneity within each stage and the dynamic nature of the climate-ecosystem relationship.

During the long glacial stages open habitats prevailed in regions and at altitudes that were not snow/ice covered. Variations in erosion intensity, due in turn to climatic variations, are reflected in the lake sediments. For example, in the record from Lake Ohrid (Vogel *et al.,* 2010) the last glacial stage was marked by variations in particle size, ranging between coarse silt to fine sand, as well as low concentrations of organic matter and calcite, which reflect low lake productivity. Moreover, the examination of the pollen content of hyena coprolites from the foothills of the Pyrenees (González-Samperiz *et al.,* 2003), which are between 50 700 and 39 900 years old, shows that varied/mosaic vegetation communities prevailed. These included juniper woodland and open habitats dominated by chenopods, grasses, etc.; the presence of more warmth-demanding trees and shrubs also raises questions about refugia, that is, loci of warmth-demanding species such as trees, from which they spread when conditions ameliorated.

The existence of refugia has been questioned, notably in the context of tropical environments (for a discussion see Haffer, 1982), which have never experienced glaciation and for which there is little evidence, but in temperate and Mediterranean environments refugia must have existed and it is a matter of speculation as to where the major tree species survived during the coldest periods of the Quaternary. This is another important facet of past, present and future ecosystem dynamics that is climate dependent and that can only be addressed through palaeoecology as a window to the past. Refugia relating to the last cold stage in the Mediterranean have been identified on the basis of pollen, macrofossil and molecular records by López de Heredia *et al.* (2007), who have examined the possibility of multiple refugia for the evergreen oak group of *Quercus suber* L., *Q. ilex* L. and *Q. coccifera* L. They suggest that refugia existed in southwestern, eastern and northern Iberia as well as the Cantabrian Mountains of southeastern Spain. Beaudouin *et al.* (2007) have summarized much of this work and have also identified the remains of *Abies, Picea and* deciduous *Quercus* from sediments in the Gulf of Lions c. 30 000 years old; this area was part of the drainage basins of the Pyreneo-Languedocian rivers and may have been linked to northeastern Spanish and southeastern French refugia/relicts. The fact that several mountain regions provided refugia during Quaternary cold stages is also considered likely by the analysis of Médail and Diadema (2009), who have examined phylogeographical evidence (i.e. molecular evidence, notably molecular markers within a species that reflect intraspecies relationships over time) for refugia in the Mediterranean region as a whole and compared such data with proposed 'hotspots' of biodiversity. Hot spots have been delineated as having high species diversity, high endemism and intense habitat modification (Médail and Quézel, 1997).

Of special significance are the interglacial stages of the last 500 000 years because they provide a record of interglacial progression similar to the present Holocene but without any human impact. Investigations of such stages provide, in principle and as far as palaeoenvironmental reconstructions allow given their limitations (for a discussion see Mannion, 1999; Wilson *et al.*, 2000), a means of investigating natural interglacial development against the concentrations of atmospheric carbon dioxide and methane, which were considerably below those of the present as indicated by concentration data from ice core records (see Lüthi *et al.*, 2008, and Loulergue *et al.*, 2008, for records of changes in atmospheric concentrations of carbon dioxide and methane during the last 800 000 years). Tzedakis (2009) and Tzedakis *et al.* (2004, 2009) have commented on the nature of interglacial stages in terms of duration and forest development, and draw attention to diversity rather than uniformity. In general, interglacials involved a vegetation succession that began with an increase in birch (*Betula*), juniper (*Juniperus*) and pine (*Pinus*). Oak (*Quercus*), elm (*Ulmus*), lime *(Tilia)* and hazel (*Corylus*) then expanded followed by hornbeam (*Carpinus*), beech (*Fagus*) and fir (*Abies*) as late invaders. However, regional variations in relation to timing and forest composition occurred as Tzedakis (2009) has reviewed in detail. Tzedakis *et al.* (2009) conclude that 'Examination of the palaeoclimate record of the past 800 000 years reveals a large diversity among interglacials in terms of their intensity, duration and internal variability, but a general theory accounting for this diversity remains elusive.' It is highly likely that such comments are just as relevant to the interglacial environments of Mediterranean mountains as they are to the mid-altitude and lowland areas in which the lake basins containing the sediments are located.

30×10^3 to 10×10^3 years ago

The Last Glacial Maximum (LGM), that is the maximum extent of ice sheets and glaciers during the last glacial stage, occurred c. 20×10^3 years ago in northern Europe and North America. The highest parts of mountains in the Mediterranean zone were ice-covered, as discussed by Hughes and Woodward (2009) and Hughes (see Chapter 3), though they point out that in the Mediterranean mountains there is evidence for maximum ice extent c. 10×10^3 years earlier with subsequent phases of glacial advance and retreat. There are a number of sites that contain pollen records of the period around and after the LGM; the synthesis by Tzedakis (2009) gives a list, described as partial, of 17 sites from Spain, France, Italy, Slovenia, Greece, Syria and Tunisia, which provide evidence for the existence of tree populations during the LGM. In general, pine (*Pinus*) and birch (*Betula*) dominate the pollen spectra of sites from the northern part of the Mediterranean Basin, while deciduous oak (*Quercus* spp.) and other deciduous species dominate the eastern and southern basin. These data could and probably do reflect climatic control of vegetation communities on the basis of latitude and possibly longitude (see also comments by Suc and Popescu, 2005) though there is the possibility that some species, such as pine (*Pinus*), are over-represented in the pollen spectra. Some of these records are from mountain regions and reinforce the idea that at least in some mountain areas there were indeed tree-covered refugia (see above) during glacial stages. Recent research, for example, based on the genetic characteristics of modern pine populations in several European mountain ranges including the Pyrenees has led Heuertz *et al.* (2010) to state that 'The core regions of the Pyrenees . . . were probably recolonized . . . by *P. uncinata* . . . from multiple glacial refugia that were well connected by pollen flow within the mountain chains'.

One example of a site indicating the existence of a refugium is Siles in the mountains of Ségura, southern Spain. This is a lake at 1320 m asl that has been investigated by Carrión (2002) using a multidisciplinary approach involving pollen, microcharcoal, spores of terrestrial plants, fungi, and non-siliceous algae, and other microfossils. The reconstructions for the 20 300 to 7400-year period are summarized in Table 2.2. These data reflect considerable environmental changes within the lake and in its catchment as the last ice age ended and they indicate that refugia extended to higher elevations than previously thought. Carrión suggests that Mediterranean and temperate tree species occupied the most sheltered habitats in a forest belt that was fragmented in comparison with that of today. Moreover, Siles is just 5–15 km from several river-cut gorges in the Parque Natural de Cazorla, Segura y Las Villas, where microclimatic conditions today are favourable to relict populations of several tree species. This study lends support to the view that the degree of shelter and/or favourable microclimatic conditions were more important than altitude in determining the location of refugia in Mediterranean mountains. Such conditions may also be more important than latitude given that Lamb *et al.* (1989) have shown, on the basis of pollen analysis of sediments from Lake Tigalmamine in the Middle Atlas Mountains at altitude 1628 m asl, that despite the more southern location in the Mediterranean Basin, sediments of the LGM were dominated by open habitat species and that tree pollen, notably oak, did not become significant until c. 14 000 years ago and persisted for c. 2000 years when a decline occurred until a further increase in the early part of the Holocene.

The decline in oak at Tigalmamine was most likely due to a decline in temperature, which is recognized as a global event during the climatically and ecologically complex period known as the Younger Dryas. This comprised a few thousand years during which ice sheets oscillated prior to a major decline as the Holocene opened (see below). There is evidence from ocean sediments as well as terrestrial sites (see Robinson *et al.*, 2006), including several in the Mediterranean mountains, for late-glacial environmental change. For example, Hughes and Woodward (2009) note the presence of late glacial moraines in the Italian Apennines, the Maritime Alps and Pyrenees. Lakes with long sequences referred to in Table 2.1. and accompanying text also record this period. For example, at Lac du Bouchet/Praclaux, 1200 m asl, steppe species expanded at the expense of forest species (Reille *et al.*, 2000), while at Ioannina, at 470 m asl in northwest Greece, Lawson *et al.* (2004) reported that woodland contracted but was not entirely replaced by open habitat species. At middle and high altitudes in the eastern Mediterranean, open habitat xerophytic (dry) herb vegetation dominated, as Denéfle *et al.* (2000) have shown based on pollen evidence from Lake Maliq (Albania) at 818 m asl in the Balkans (see also Willis, 1994). At Lake Dolgoto at 2310 m asl in the Pirin Mountains of

Table 2.2 Environmental change at Lake Siles, Ségura, southern Spain, during the period 20 300 to 7400 years BP (based on Carrión, 2002)

Age (cal. years BP)	Pollen and vegetation	Other
11 900–7400	Early Holocene expansion of pine, probably *Pinus nigra*. A sudden expansion indicates a rapidly rising treeline as temperatures ameliorate. Increases in the pollen of thermophilous species, e.g. *Phillyrea, Olea*, Ericaceae and evergreen oak (*Quercus*) Between 10 100 and 7400 years BP pine continues to dominate but *P. pinaster* replaces *P. nigra*. There are slight increases in *Juniperus*, Poaceae, *Artemisia, Ephedra nebrodensis*, Chenopodiaceae and other heliophytes. After 8100 years BP *Betula, Fraxinus*, deciduous *Quercus, Corylus* and *Acer* begin to increase	10 100–8100 years BP shallowing of the lake occurred with two periods of drying. This also reflects warming
20 300–11 900	Steppe grassland comprising Poaceae with *Artemisia, Ephedra nebrodensis* and a scattering of pine and juniper (*Juniperus*). Other open-habitat species include Chenopodiaceae, Caryophyllaceae, Brassicaceae and *Helianthemum* By 12 000 years BP *Artemisia, E. nebrodensis* and Chenopodiaceae had declined as juniper expanded gradually within the steppe grassland. *P. pinaster* and evergreen *Quercus* expanded Persistent occurrences of the warmth-demanding *Acer, Taxus, Arbutus, Buxus, Salix, Ulmus, Phillyrea, Pistacia* and *Olea* indicate the proximity of glacial refugia of temperate and Mediterranean trees	Until 10 100 years BP colonization by thermophilous microphytes (e.g. *Closterium–Zygnema, Debarya–Potamogeton*) coincided with a reduction in cryoclastic scree content and particle size. This may have been due to slope stabilization as vegetation colonised Productivity was low during the earlier stage but increased with ameliorating temperatures

Bulgaria, Stefanova and Ammann (2003) have shown that the vegetation during the period 11 000–10 200 ^{14}C years BP (radiocarbon age before present) was open and dominated by *Artemisia* and Chenopodiaceae and that soil erosion was occurring in the lake catchment. Relatively harsh conditions thus prevailed in these mountains prior to Holocene amelioration. What these records do not reveal is whether the pre-11 000-year period experienced higher temperatures and that there was a late-glacial climatic regression following an amelioration as was the case in many other regions.

However, there is evidence for changing hydrological conditions with relatively low lake levels (and probably higher salinities) and drier conditions being characteristic of this late-glacial cold period, as Roberts and Reed (2009) have indicated on the basis of palaeoenvironmental research in Italy and Turkey, though the sites described are generally at low altitude. However, Hajar *et al.* (2008) have recorded the Younger Dryas in sediments from the Aammiq wetland at 865 m asl in the Bekaa Valley, Lebanon. Here an expansion of cedar and oak was halted c. 12 200 years ago when a chenopod-rich steppe developed due to cooling. Similarly, at the Lago di Mezzano at 452 m asl in central Italy, forest expansion was curtailed c. 12 000 years ago (Ramrath *et al.*, 2000). The patterns indicated by these sites may reflect an east-west gradient, an altitudinal gradient, or both.

Between the last glacial maximum and the beginning of the Holocene, Mediterranean mountain environments below the snow line were most probably composed of mosaics of vegetation types. Open habitat shrub and grass communities dominated exposed slopes, with scattered trees on less exposed slopes and mixed woodlands in sheltered valleys. Shifts in community locations and composition no doubt occurred following temperature changes that crossed thresholds of tolerance for some but not all species. It is likely that earlier glacial periods were similarly characterized. The late glacial was a complex period climatically with evidence for widespread forest development, which was interrupted as steppe returned and/or forest declined during the Younger Dryas for a few thousand years prior to the start of the Holocene.

2.4 The Holocene

There is a considerable body of evidence for environmental change during the Holocene – the last 10 000 years – in the Mediterranean mountains, especially from lake sediments. The literature is vast and some of it has recently been reviewed by Tzedakis (2009) and Roberts and Reed (2009). Broadly, a distinction can be made between the early and late Holocene, which reflects climatic change and intensifying human impact due to the initiation of agriculture. This, plus a transitional zone, has been proposed by Jalut *et al.* (2009) whose synthesis is based on the published pollen analysis of 23 sites within the Mediterranean Basin, though Tigalmamine is the only site in the southern basin; three sites are from the east and 19 are from the north. A summary is given in Table 2.3. A significant implication of this synthesis is that aridification began as a response to natural interglacial climatic change and was later amplified by human activity.

Another factor that affected Holocene vegetation development was fire. Moreover, there is considerable debate as to how significant fire was in comparison with climate as a driver of early Holocene forest change, notably the development of evergreen oak (*Quercus ilex*) communities beginning c. 8500 cal. years BP (calibrated years before present). Addressing this issue, Colombaroli *et al.* (2009) examined the pollen and charcoal profiles in a number of lake sites in the Adriatic region and concluded that climatic change was the primary driving force on vegetation

Table 2.3   A summary of Holocene environmental change (based on Jalut *et al.*, 2009)

Age (cal. years BP)	Division	Climate	Vegetation, etc.
5500 to present	Upper Holocene	Aridification	A decline of deciduous broad-leaved trees and spread of evergreen sclerophyllous taxa. This was prompted by climatic characteristics and probably accelerated as human activity intensified
7000–5500	Transition phase	Decrease in insolation; development of an atmospheric system similar to that of today	
11 500–7000	Lower Holocene	Humid with dry episodes	Spread of deciduous broadleaved trees

change overall. They also showed that this was the case during later periods when Mesolithic and Neolithic activity was occurring. It is also of interest that they identified a north-south gradient; that is, at the northern sites in Tuscany and Croatia *Quercus ilex* replaced deciduous forest species as conditions became drier; in contrast at the more southerly sites in Sicily *Q. ilex* replaced maquis or steppe as conditions became wetter. This is because in wetter conditions, as in the early Holocene of the north Adriatic region, deciduous species can outcompete *Quercus ilex* while under drier conditions, as in the south, *Quercus ilex* is outcompeted by shrubs and herb species. The sites investigated in this study are all at low altitude and how far the results and inferences can be extrapolated in relation to the rest of the Mediterranean Basin or its mountains requires further research. Nevertheless, it reflects the complexity of climate-vegetation-fire relationships, which are just as important in Mediterranean mountains. A further point of interest is the contrast between earlier interglacial development and the Holocene. Tzedakis (2009) has reviewed this in detail. He points out that sclerophyll species underwent a greater expansion during the early part of the last interglacial than in the Holocene, which he attributes to increased summer insolation in the interglacial. In comparison hazel (*Corylus*) was more significant in the early Holocene in the western Mediterranean, with oak (*Quercus*) in the east. Moreover, interglacial vegetation successions were more diversified than those of the Holocene and, of course, the latter half of the Holocene was substantially affected by human activity not least by the spread of the olive (*Olea*) and the creation of pasture and arable land during Neolithic times.

Notwithstanding these generalizations, examples of results from a few upland sites illustrate specific vegetation/environmental changes. Table 2.4 summarizes the environmental changes during the last 18 000 years at Tigalmamine in the

Table 2.4 Examples of late Quaternary vegetation change

Date	Characteristics	Date	Characteristics
Site: Tigalmamine, Morocco, 1626 m asl (Lamb *et al.*, 1989)		Site: Aammiq, Bekaa Valley, Lebanon, 865 m asl (Hajar *et al.*, 2008)	
450	Increase in *Cedrus atlantica*	3400–1000	Evergreen and deciduous oaks with cedars
2250	Human impact: forest degradation	6100–3400	Sedimentary process disturbed; aridity/human?
4000	*Cedrus atlantica* arrives	11 500–6100	Open oak forest locally with cedars on Barouk Mt
8500	Oak (*Quercus*) forest develops	By 11 000	Aammiq marsh develops
12 000–8500	Herb-rich grassland predominates	12 000–11 000	Chenopodiaceae steppe with scattered cedars
14 000–12 000	Scattered oaks invade grassland	pre-12 200	Expansion of cedars and deciduous oaks
18 000–14 000	Herb-rich grassland predominates		

Middle Atlas Mountains of Morocco and shows that herb-rich grassland was replaced with oak forest, with cedar being a relatively late invader. At Siles (see Table 2.2), at 1320 m asl in the Segura Mountains of Spain, pine spread into early Holocene grassland rather than oak and there is no evidence for cedar, which was confined to the southern and eastern parts of the Mediterranean Basin. Indeed there is evidence for the presence of cedar (*Cedrus libani*) in the Aammiq wetland, Bekaa valley, pre-12 200 years BP as shown in Table 2.4. Here in the surrounding mountains oak forests and cedars dominate the Holocene.

During the later Holocene, post-5000 years BP, human impact is evident throughout the region. Its toll on forests plus natural processes of aridization resulted in forest fragmentation and the dominance of shrubs, which are typical of the region today. The rise and fall of many of the world's great civilizations caused vegetation change and shaped the land to create anthropogenic landscapes that are today occupied by some 450 million people.

2.5 Conclusion

The Quaternary (and earlier) history of the Mediterranean mountains and the Mediterranean Basin in general is complex. Evidence derives from a variety of marine and terrestrial sources and the most abundant line of evidence is pollen analysis. There is considerable information available from the northern part of the basin, which includes several long lake sequences spanning more than one glacial-interglacial cycle, limited sources in the east but relatively little from the south. It is also important to recognize that many sites referred to are not in upland areas and the results can only be extrapolated. Consequently, reconstructions are spatially rather than temporally constrained. Nevertheless some patterns are discernible, with domination by climatic change as a driving force in ecosystem change, which is evident during glacial-interglacial cycles. It is also possible to identify spatial gradients notably in terms of ecosystem characteristics between west and east as well as north and south. Undoubtedly, the latter part of the Holocene has no counterpart in earlier interglacial stages given the heavy footprint of humanity, which has created and shaped landscapes for the last ten millennia.

References

References marked as bold are key references.

Anderson, D.E., Goudie, A.S. and Parker, A.G. (2007) *Global Environments Through the Quaternary.* Oxford: Oxford University Press.

Babault, J., Loget, N., Van Den Driessche, J., Castelltort, S., Bonnet, S. and Davy, P. (2006) Did the Ebro basin connect to the Mediterranean before the Messinian salinity crisis? *Geomorphology* 81:155–165.

Beaudouin, C., Gwénaël., J., Suc, J.-C., Berné, S. and Escarguel, G. (2007) Vegetation dynamics in southern France during the last 30 ky BP in the light of marine palynology. *Quaternary Science Reviews* 26:1037–1054.

Béthoux, J.-P. and Pierre, C. (1999) Mediterranean functioning and sapropel formation: respective influences of climate and hydrological changes in the Atlantic and the Mediterranean. *Marine Geology* 153:29–39.

Blanc, P.-L. (2002) The opening of the Plio-Quaternary Gibraltar strait: assessing the size of a cataclysm. *Geodinamica Acta* 15:303–317.

Butler, R.W.H., McClelland, E. and Jones, R.E. (1999) Calibrating the duration and timing of the Messinian salinity crisis in the Mediterranean: linked tectonoclimatic signals in thrust-top basins in Sicily. *Journal of the Geological Society, London* 156:827–835.

Carrión, J.S. (2002) Patterns and processes of Late Quaternary environmental change in a montane region of southwestern Europe. *Quaternary Science Reviews* 21:2047–2066.

Cheddadi, R., de Beaulieu, J.-L., Jouzels, J. *et al.* (2005) Similarity of vegetation dynamics during interglacial periods. *Proceedings of the National Academy of Sciences of the USA* 102:13939–13943.

CIESM (2008) The Messinian salinity crisis from mega-deposits to microbiology – A consensus report. CIESM workshops No. 33. Monaco: CIESM.

Clauzon, G., Suc, J.-P., Gautier, F., Berger, A. and Loutre, M.-F. (1996) Alternate interpretation of the Messinian salinity crisis: Controversy resolved? *Geology* 24:363–366.

Colombaroli, D., Tinner, W., van Leeuwen, J. *et al.* (2009) Response of broadleaved evergreen Mediterranean forest vegetation to fire disturbance during the Holocene: insights from the peri-Adriatic region. *Journal of Biogeography* 36:314–326.

Denéfle, M., Lézine, A.-M., Fouache, E. and Dufaure, J.-J. (2000) A 12,000-year pollen record from Lake Maliq, Albania. *Quaternary Research* 54:423–432.

Duggen, S., Hoernie, K., van den Bogaard, P., Rupke, L. and Phipps Morgan, J. (2003) Deep roots of the Messinian salinity crisis. *Nature* 422:602–605.

Emeis, K.-C. and Weissert, H. (2009) Tethyan–Mediterranean organic carbon-rich sediments from Mesozoic black shales to sapropels. *Sedimentology* 56:247–266.

Fauquette, S., Suc, J-P., Bertini, A. *et al.* (2006) How much did climate force the Messinian salinity crisis? Quantified climatic conditions from pollen records in the Mediterranean region. *Palaeogeography, Palaeoclimatology, Palaeoecology* 238:281–301.

Favre, E., François, L., Fluteau, F., Cheddadi, R., Thévenod, L. and Suc, J.-P. (2007) Messinian vegetation maps of the Mediterranean region using models and interpolated pollen data. *Geobios* 40:433–443.

Follieri, M., Magri, D. and Sadori, L. (1988) 250,000-year pollen record from Valle di Castiglione (Roma). *Pollen et Spores* 30:329–356.

Gargani, J. and Rigollet, C. (2007) Mediterranean Sea level variations during the Messinian salinity crisis. *Geophysical Research Letters* 34:L10405; doi: 1029/2007GL029885.

Gibbard, P.L., Head, M.J. and Walker, M.J.C. (2009) Formal ratification of the Quaternary System/Period and the Pleistocene Series/Epoch with a base at 2.58 Ma. *Journal of Quaternary Science* 24:96–102.

González-Samperiz, P., Montes, L. and Utrilla, P. (2003) Pollen in hyena coprolites from Gabasa Cave (northern Spain). *Review of Palaeobotany and Palynology* 126:7–15.

Haffer, J. (1982) General aspects of the refuge theory. In Prance, G. (ed.), *Biological Diversification in the Tropics*. New York: Columbia University Press, pp. 6–24.

Hajar, L., Khater, C. and Cheddadi, R. (2008) Vegetation changes during the late Pleistocene and Holocene in Lebanon: a pollen record from the Bekaa Valley. *The Holocene* 18:1089–1099.

Haug, G.H., Ganopolski, A., Sigman, D.M. *et al.* (2005) North Pacific seasonality and the glaciation of North America 2.7 million years ago. *Nature* 433:821–825.

Heuertz, M., Teufel, J., González-Martinez, S.C. *et al.* (2010) Geography determines genetic relationships between species of mountain pine (*Pinus mugo* complex) in western Europe. *Journal of Biogeography* 37:541–556.

Hilgen, F.J. (1991) Astronomical calibration of the Gauss to Matuyama sapropels in the Mediterranean and implications for the geomagnetic polarity timescale. *Earth and Planetary Science Letters* 104:226–244.

Horowitz, A. (1989) Continuous pollen diagrams for the last 3.5Ma from Israel: Vegetation, climate and correlation with the oxygen isotope record. *Palaeogeography, Palaeoclimatology, Palaeoecology* 7:63–78.

Hughes, P. and Woodward, J. (2009) Glacial and periglacial environments. In: Woodward, J. (ed.), *The Physical Geography of the Mediterranean*. Oxford: Oxford University Press, pp. 353–383.

Jalut, G., Dedoubat, J.J., Fontugne, M. and Otto, T. (2009) Holocene circum-Mediterranean vegetation changes: Climate forcing and human impact. *Quaternary International* 200:4–18.

Joannin, S., Quillévéré, F., Suc, J.-P., Lécuyer, C. and Martineau, F. (2007a) Early Pleistocene climate changes in the central Mediterranean region as inferred from integrated pollen and planktonic foraminiferal stable isotope analyses. *Quaternary Research* 67:264–274.

Joannin, S., Cornée, J.-J., Moissette, P. *et al.* (2007b) Changes in vegetation and marine environments in the eastern Mediterranean (Rhodes, Greece) during the Early and Middle Pleistocene. *Journal of the Geological Society of London* 164:1119–1131.

Joannin, S., Ciaranfi, N. and Stefanelli, S. (2008) Vegetation changes during the late Early Pleistocene at Montalbano Jonico (Province of Matera, southern Italy) based on pollen analysis. *Palaeogeography, Palaeoclimatology, Palaeoecology* 270:92–101.

Jolivet, L., Augier, R., Robin, C., Suc, J.-P. and Rouchy, J.-M. (2006) Lithospheric-scale geodynamic context of the Messinian salinity crisis. *Sedimentary Geology* 188–189:9–33.

Krijgsman, W., Hilgen, F.J., Raffi, I., Sierro, F.J. and Wilson, D.S. (1999) Chronology, causes and progression of the Messinian salinity crisis. *Nature* 400:652–655.

Kroon, D., Alexander, I., Little, M. *et al.* (1998) Oxygen isotope and sapropel stratigraphy in the eastern Mediterranean during the last 3.2 million years. *Proceedings of the Ocean Drilling Program, Scientific Results* 160:181–189.

Lamb, H.F., Eicher, U. and Switsur, V.R. (1989) An 18,000-year record of vegetation, lake-level and climatic change from Tigalmamine, Middle Atlas, Morocco. *Journal of Biogeography* 16:65–74.

Lawson, I., Frogley, M., Bryant, C., Preece, R., and Tzedakis, P. (2004) The Late glacial and Holocene environmental history of the Ioannina basin, north-west Greece. *Quaternary Science Reviews* 23:1599–1625.

López de Heredia, U., Carrión, J.S., Jiménez, P., Collada, C. and Gil, L. (2007) Molecular and palaeoecological evidence for multiple glacial refugia for evergreen oaks on the Iberian Peninsula. *Journal of Biogeography* 34:1505–1517.

Loulergue. L., Schilt, A., Spahni, R. *et al.* (2008) Orbital and millennial-scale features of atmospheric CH4 over the past 800,000 years. *Nature* 453:383–386.

Lüthi, D., Le Floch, M., Bereiter, B. *et al.* (2008) High-resolution carbon dioxide concentration record 650,000–800,000 years before present. *Nature* 453:379–382.

Magri, D. (1989) Interpreting long-term exponential growth of plant populations in a 250,000-year pollen record from Valle di Castiglione (Roma). *New Phytologist* 112:123–128.

Mannion, A.M. (1999) *Natural Environmental Change*. London: Routledge.

Mannion, A.M. (2006) *Carbon and Its Domestication*. Dordrecht: Springer.

Mannion, A.M. (2008) The Tertiary-Quaternary environmental history of the Mediterranean basin: The background to Mediterranean Island environments. In Vogiatzakis, I.N., Pungetti, G. and Mannion, A.M. (eds), *Mediterranean Island Landscapes. Natural and Cultural Approaches*. Dordrecht: Springer, pp. 15–35.

Médail, F. and Diadema, F. (2009) Glacial refugia influence plant diversity patterns in the Mediterranean Basin. *Journal of Biogeography* 36:1333–1345.

Médail, F. and Quézel, P. (1997) Hot-spots analysis for conservation of plant biodiversity in the Mediterranean Basin. *Annals of the Missouri Botanical Garden* 84:112–127.

Meyers, P.A. (2006) Paleoceanographic and paleoclimatic similarities between Mediterranean sapropels and Cretaceous black shales. *Palaeogeography, Palaeoclimatology, Palaeoecology* 35:305–320.

Meyers, P.A. and Arnaboldi, M. (2008) Paleoceanographic implications of nitrogen and organic carbon isotopic excursions in mid-Pleistocene sapropels from the Tyrrhenian and Levantine Basins, Mediterranean Sea. *Palaeogeography, Palaeoclimatology, Palaeoecology* 266:112–118.

Mijarra, J.M.P., Barron, E., Manzaneque, F.G. and Morla, C. (2009) Floristic changes in the Iberian Peninsula and Balearic Islands (south-west Europe) during the Cenozoic. *Journal of Biogeography* 36:2025–2043.

Mommersteeg, H.J.P.M., Loutre, M.F., Young, R., Wijmstra, T.A. and Hooghiemstra, H. (1995) Orbital forced frequencies in the 975,000 year pollen record from Tenaghi Philippon (Greece). *Climate Dynamics* 11:4–24.

Nebout, N.C., Londeix, L., Baudin, F., Turon, J.-L., von Grafenstein, R. and Zahn, R. (1999) Quaternary marine and continental paleoenvironments in the western Mediterranean (site 976, Alboran Sea): palynological evidence. *Proceedings of the Ocean Drilling Program, Scientific Results* 161:457–468.

Okuda, M., Yasuda, Y. and Setoguchi, T. (2001) Middle to late Pleistocene vegetation history and climatic changes at Lake Kopais, southeast Greece. *Boreas* 30:73–82.

Popescu, S.-M., Suc, J.-P. and Loutre, M.-F. (2006) Early Pliocene vegetation changes forced by eccentricity-precession. Example from Southwestern Romania. *Palaeogeography, Palaeoclimatology, Palaeoecology* 238:340–348.

Popescu, S.-M., Biltekin, D., Winter, H. *et al.* (2010) Pliocene and Lower Pleistocene vegetation and climate changes at the European scale: Long pollen records and climatostratigraphy. *Quaternary International* 219:152–167.

Rackham, O. (2008) Holocene history of Mediterranean island landscapes. In Vogiatzakis, I.N., Pungetti, G. and Mannion, A.M. (eds), *Mediterranean Island Landscapes. Natural and Cultural Approaches.* Dordrecht: Springer, pp. 36–60.

Ramrath, A., Sadori, L. and Negendank, J.F.W. (2000) Sediments from Lago di Mezzano, central Italy: a record of Lateglacial/Holocene climatic variations and anthropogenic impact. *The Holocene* 10:87–95.

Reille, M. and de Beaulieu, J.-L. (1995) Long Pleistocene pollen records from the Praclaux Crater, south-central France. *Quaternary Research* 44:205–215.

Reille, M., Andrieu, V., de Beaulieu, J.-L., Guenet, P. and Goeury, C. (1988) A long pollen record from Lac du Bouchet, Massif Central, France for the period 325 to 100 ka (OIS 9c to OIS 5e). *Quaternary Science Reviews* 17:1107–1123.

Reille, M., De Beaulieu, J.-L., Svoboda, H., Andrieu-Ponel, V. and Goevry, C. (2000) Pollen analytical biostratigraphy of the last five climatic cycles from a long continental sequence from the Velay region (Massif Central, France). *Journal of Quaternary Science* 15: 665–685.

Roberts, N. and Reed, J. (2009) Lakes, wetlands and Holocene environmental change. In: Woodward, J. (ed.), *The Physical Geography of the Mediterranean.* Oxford: Oxford University Press, pp. 255–286.

Robinson, S.A., Black, S., Sellwood, B.W. and Valdes, P.J. (2006) A review of palaeoclimates and palaeoenvironments in the Levant and eastern Mediterranean from 25000 to 5000 years BP: setting the environmental background for the evolution of human civilisation. *Quaternary Science Reviews* 25:1517–1541.

Ryan, W.F.B. (2009) Decoding the Mediterranean salinity crisis. *Sedimentology* 56:95–136.

Shackleton, N.J., Backman, J., Zimmerman, H. *et al.* (1984) Oxygen isotope calibration of the onset of ice rafting and history of glaciation in the North Atlantic region. *Nature* 307:620–623.

Stefanova, I. and Ammann, B. (2003) Lateglacial and Holocene vegetation belts in the Pirin Mountains (southwestern Bulgaria). *The Holocene* 13:97–107.

Subally, D., Bilodeau, G., Tamrat, E., Ferry, S., Debard, E. and Hillaire-Marcel, C. (1999) Cyclic climatic records during the Olduvai Subchron (uppermost Pliocene) on Zakynthos island (Ionian Sea). *Geobios* 32:793–803.

Subally, D. and Quézel, P. (2002) Glacial or interglacial: *Artemisia*, a plant indicator with dual responses. *Review of Palaeobotany and Palynology* 120:123–130.

Suc, J.-P. (1984) Origin and evolution of the Mediterranean vegetation and climate in Europe. *Nature* 307:429–432.

Suc, J.-P., Bertini, A., Combourieu-Nebout, N. *et al.* (1995) Structure of West Mediterranean vegetation and climate since 5.3 Ma. *Acta Zoologica Cracova* 38:3–16.

Suc, J.-P. and Popescu, S.-M. (2005) Pollen records and climatic cycles in the North Mediterranean region since 2.7 Ma. In: Head, M.J. and Gibbard, P.L. (eds) *Early Middle Pleistocene Transitions: The Land–Ocean Evidence*. London: Geological Society Special Publications 247, pp. 147–158.

Suc, J.-P., Combourieu-Nebout, N., Seret, G. *et al.* (2010) The Crotone series: a synthesis and
new data. *Quaternary International* 219:121–133.

Tzedakis, P.C. (1999) The last climatic cycle at Kopais, central Greece. *Journal of the Geological
Society of London* 156:425–434.

Tzedakis, P.C. (2009) Cenozoic climate and vegetation change. In: Woodward, J. (ed.), *The Physical Geography of the Mediterranean*. Oxford: Oxford University Press, pp. 89–137.

Tzedakis, P.C., Frogley, M.R. and Heaton, T.H.E. (2003) Last interglacial conditions in southern Europe: evidence from Ioannina, northwest Greece. *Global and Planetary Change*
36:157–170.

Tzedakis, P.C., Roucoux, K.H., de Abreu, L., Shackleton, N.J. (2004) The duration of forest
stages in southern Europe and interglacial climate variability. *Science* 306:2231–2235.

Tzedakis, P.C., Hooghiemstra, H. and Pälike, H. (2006) The last 1.35 million years at Tenaghi
Philippon: revised chronostratigraphy and long-term vegetation trends. *Quaternary Science
Reviews* 25:3416–3430.

Tzedakis, P.C., Raynaud, D.M, McManus, J.F., Berger, A., Brovkin, V. and Keifer, T. (2009)
Interglacial diversity. *Nature Geoscience* 2:751–755.

Van Couvering, J.A. (ed.) (2004) *The Pleistocene Boundary and the Beginning of the Quaternary*.
Cambridge: Cambridge University Press.

Vogel, H., Wagner, B., Zanchetta, G., Sulpizio, R. and Rose, P. (2010) A paleoclimate record
with tephrochronological age control for the last glacial-interglacial cycle from Lake Ohrid,
Albania and Macedonia. *Journal of Paleolimnology* 44:295–310.

Wagner, B., Lotter, A.F., Nowaczyk, N. *et al.* (2009) A 40,000-year record of environmental change from ancient Lake Ohrid (Albania and Macedonia). *Journal of Paleolimnology*
41:407–430.

Warny, S.A., Bart, P.J. and Suc, J.-P. (2003) Timing and progression of climatic, tectonic and
glacioeustatic influences on the Messinian Salinity Crisis. *Palaeogeography, Palaeoclimatology, Palaeoecology* 202:59–66.

Willett, S.D., Schlunegger, F. and Picotti, V. (2006) Messinian climate change and erosional destruction of the central European Alps. *Geology* 34:613–616.

Willis, K.J. (1994) The vegetational history of the Balkans. *Quaternary Science Reviews*
13:769–788.

Wilson, R.C.L., Drury, S.A. and Chapman, J.L. (2000) *The Great Ice Age*. London: Routledge.

Zagwijn, W.H. (1960) Aspects of the Pliocene and early Pleistocene vegetation in the Nether-
lands. *Mededelingen Geologische Stichting* Serie C 3 (5), 78 pp.

3

Glacial history

Philip D. Hughes

3.1   Introduction

Many of the Mediterranean mountains supported glaciers during the Pleistocene –
from Morocco in the west to the Lebanon in the east (Hughes *et al.*, 2006a; Hughes
and Woodward, 2009; Hughes, 2011) and some glaciers and ice patches still survive
today (Figure 3.1). In the Maritime Alps, north of the Côte d'Azur, Pleistocene ice
caps were contiguous with the main Alpine ice sheet, which covered a total area
of 126 000 km^2 during the last cold stage (Ehlers, 1996). In this chapter, only the
southernmost parts of the main Alpine chain that border the Mediterranean Sea are
considered. Ice caps also formed over the mountains of northwest Iberia (Vieira,
2007; Cowton *et al.*, 2009) and also over large areas of the western Balkans (Hughes
et al., 2011a). Glaciers even formed as far south as Crete during the Pleistocene
(Fabre and Maire, 1983). The consequences of these glaciations are clearly seen
in the Mediterranean mountains where glaciers have shaped the landscape to form
cirques, U-shaped valleys, arêtes, roches moutonnées, glacial lakes and moraines.
Today there are still some glaciers in the Mediterranean mountains. Most of these
are restricted to the highest mountains such as the Pyrenees, the Maritime Alps and
the mountains of Turkey where the highest summits exceed 3000 m asl. However,
several glaciers also exist in lower mountain areas, such as central Italy (D'Orefice
et al., 2000) and in Montenegro and Albania (Hughes, 2007, 2009).

There is a long history of glacial research in the Mediterranean mountains. Jovan
Cvijić was one of the first geographers to note and map glacial landforms in Europe
and worked extensively in the Balkans in the late nineteenth and early twentieth
centuries (see, e.g., Cvijić, 1896, 1908, 1909, 1913). Albrecht Penck was another
early scholar who worked in the Mediterranean mountains. His research covered
a range of different areas including the Pyrenees (France/Spain) and the Dinaric
Alps (Balkans) (Penck, 1900). Hughes *et al.* (2006a) identified the first phase in
the development of the subject as the 'Pioneer Phase'. This phase is characterized
by initial observations of glacial features in the mountains and largely involves
descriptive accounts of features such as cirques, U-shaped valleys, moraines and

Mediterranean Mountain Environments, First Edition. Edited by Ioannis N. Vogiatzakis.
© 2012 John Wiley & Sons, Ltd. Published 2012 by John Wiley & Sons, Ltd.

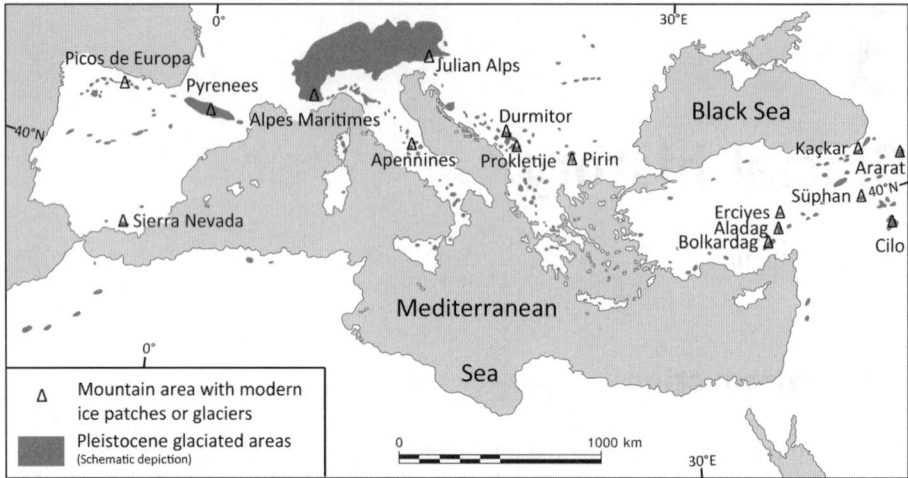

Figure 3.1 The extent of past Pleistocene glaciers in the Mediterranean mountains showing the locations of surviving (or recently extinct) modern-day glaciers and ice patches in the twenty-first century. Reproduced from Hughes (2011), with permission

erratics. The pioneer phase continued for decades through the twentieth century and glacial research in the Mediterranean was particularly active in the inter-war period between 1920 and 1940. For example, numerous papers were written on the glacial landforms of the High Atlas of Morocco during the 1920s (Martonne, 1924; Celerier and Charton, 1922; 1923) whilst several papers on the glacial landforms of Greece and the Balkans were also published in the inter-war period (Maull, 1921; Louis, 1926, 1930; Nowack, 1929; Sestini, 1933).

Some early glacial scholars, such as Cvijić (1913) and Sawicki (1911), went beyond this pioneer phase and mapped glacial landforms on geomorphological maps. Hughes *et al.* (2006a) identified this development as the 'Mapping Phase'. However, many of the early geomorphological maps were basic and presumed that all of the glacial features formed during one ice age event. Furthermore, the transition between pioneer phase research, whereby researchers may have depicted glacial features on simple geomorphological maps, and mapping phase research, whereby researchers produced 'more detailed' geomorphological maps, is not always clear. Scale and coverage are important criteria, in addition to detail, and papers presenting glacial geomorphological maps covering the whole, or large parts, of a massif at scales of 1:100 000 or less can be considered to be in the mapping phase. The pioneer phase was largely complete for most areas of the Mediterranean by the end of the 1930s yet the mapping phase took much longer to complete, and in some areas is yet to be completed. Also, the level of detail provided in published works of the mapping phase is highly variable, with some studies simply marking moraines and cirques whilst others identify a range of glacial features and also recognize different phases of glaciation and subdivide different landforms within a chronostratigraphical framework. In Bruno Messerli's classic review of glaciation (Messerli, 1967) many areas still remained in the pioneer phase although the volume of research on the glacial history of the Mediterranean was very large, with 364 publications cited in Messerli's review. However, whilst Messerli's paper could be regarded as representing the 'watershed in pioneer phase research' in the Mediterranean mountains (Hughes *et al.*, 2006a, p. 335), the last three decades of the twentieth century was a period of stagnation in the development of ideas with few papers advancing knowledge beyond the mapping phase.

The final phase in the evolution of glacial research in the Mediterranean mountains was termed the 'advanced phase' by Hughes *et al.* (2006a) and is characterized by the application of radiometric dating, and the development of detailed geochronologies combined with modern sedimentological and stratigraphical analyses. The earliest contributions to this phase applied radiocarbon dating to glacial lake sediments. Often this research was undertaken not by glacial geomorphologists but by palaeoecologists, such as Janssen and Woldringh (1981) in the Serra da Estrela, Portugal, Mardones and Jalut (1983) and Andrieu *et al.* (1988) in the Pyrenees, and Allen *et al.* (1996) in the Sanabria National Park, Spain. However, in the 1990s radiocarbon dating was utilized directly by scholars such as Giraudi and Frezzotti (1997) for understanding the timing of glaciations. The number of studies examining the geochronology of glaciations in the Mediterranean mountains has accelerated dramatically in the first decade of the twenty-first century, especially with the development of cosmogenic nuclide exposure dating. Many new papers are now appearing every year, whereas just a decade ago glacial research in the region was largely dormant. Thus, it would seem that our knowledge of glacial history of the Mediterranean region is rapidly reaching an advanced phase. However, that is not to say that our knowledge is complete. New evidence on the timing of glacial advances in different parts of the Mediterranean is presenting new problems and questions, and this is likely to spur a new era of prolific research in this field.

3.2 Pleistocene glaciations in the Mediterranean mountains

The mountains surrounding the Mediterranean basin have been glaciated on several occasions through the Quaternary (Hughes *et al.*, 2006a). The geological and geomorphological record of glaciation in these mountains provides valuable information on past climates, since glaciers are closely related to atmospheric air temperatures and moisture supply. Ice has sculpted and shaped pyramidal peaks, arêtes, cirques and glacial troughs and produced moraines and glacial lakes in areas as diverse as Morocco and Montenegro. Glaciers have also affected other processes and had a profound impact on fluvial systems down-valley (Woodward *et al.*, 2008) and, in limestone massifs, left a major imprint on karst geomorphology (Hughes *et al.*, 2007a).

In many areas, glacial deposits record glaciations during multiple Pleistocene cold stages. The oldest and most extensive glaciations occurred during the Middle Pleistocene, *before* the last interglacial (c. 125–110 ka)* and the last cold stage of the Late Pleistocene (c. 110–11.7 ka) – see van Kolfschoten *et al.* (2003) and Walker *et al.* (2009) for chronostratigraphical definitions of the Late Pleistocene. In northwest Spain Middle Pleistocene glaciations have been shown using cosmogenic nuclide analyses (Vidal-Romaní *et al.*, 1999; Fernandez Mosquera *et al.*, 2000; Vidal-Romaní and Fernández-Mosquera, 2006), whilst in Greece Middle Pleistocene glacial deposits have been recognized using uranium-series dating (Hughes *et al.*, 2006b). Elsewhere, in places such as Corsica, the central Balkans and parts of Turkey, cosmogenic nuclide analyses suggest that the most extensive recorded glaciation occurred during the last cold stage of the Pleistocene and no evidence of earlier glaciations has been reported (Kuhlemann *et al.*, 2008, 2009; Akçar *et al.*, 2007; 2008). The timing of glacier advances may have been asynchronous across the Mediterranean, and the extent of glaciations also appears to have varied between regions during different Pleistocene cold stages (Hughes and Woodward, 2008). A detailed review of the current state of knowledge of Pleistocene glaciations in the Mediterranean is provided below, region by region.

3.2.1 Atlas Mountains, North Africa

The Atlas Mountains contain the highest mountains of North Africa and some of the highest mountains surrounding the Mediterranean Basin, several of which exceed 4000 m asl. As a consequence of the high elevation of the High Atlas, large glaciers formed during the Pleistocene especially in the High Atlas. However, no glaciers survive today (Hughes *et al.*, 2011b).

The largest glaciers formed in the Toubkal massif (4167 m asl), where valley glaciers emanated from a central ice field that formed between the two highest summits, Toubkal (4167 m asl) and Ouanoukrim (4067 m asl). The northern outlet glacier draining this ice field extended nearly 10 km to an altitude of just 2000 m, about 1 km down-valley of the shrine of Sidi Chamarouch. Valley glaciers also formed on the northern slopes of Aksoual and Bou Iguenouane. Well-defined cirques are present in the upper catchments around Toubkal and on the northern slopes of Aksoual and Bou Iguenouane, and these cirques contain moraines (Figures 3.2 and 3.3). In some glacial valleys, large landslides are closely associated with moraine complexes (Figure 3.4). Beryllium-10 exposure ages suggest that glaciers formed in the High Atlas during the Younger Dryas – a well-known cold reversal in the North Atlantic region. A lower set of moraines have yielded [10]Be exposure ages close to the global Last Glacial Maximum (LGM) whilst the lowest

---

*Note: ages are abbreviated in thousands of years (e.g. 40 500 years = 40.5 ka). Radiocarbon ages are presented as 40 ka ^{14}C BP, with BP standing for 'before present' (AD 1950). Calibrated radiocarbon ages are denoted as 40 cal. ka.

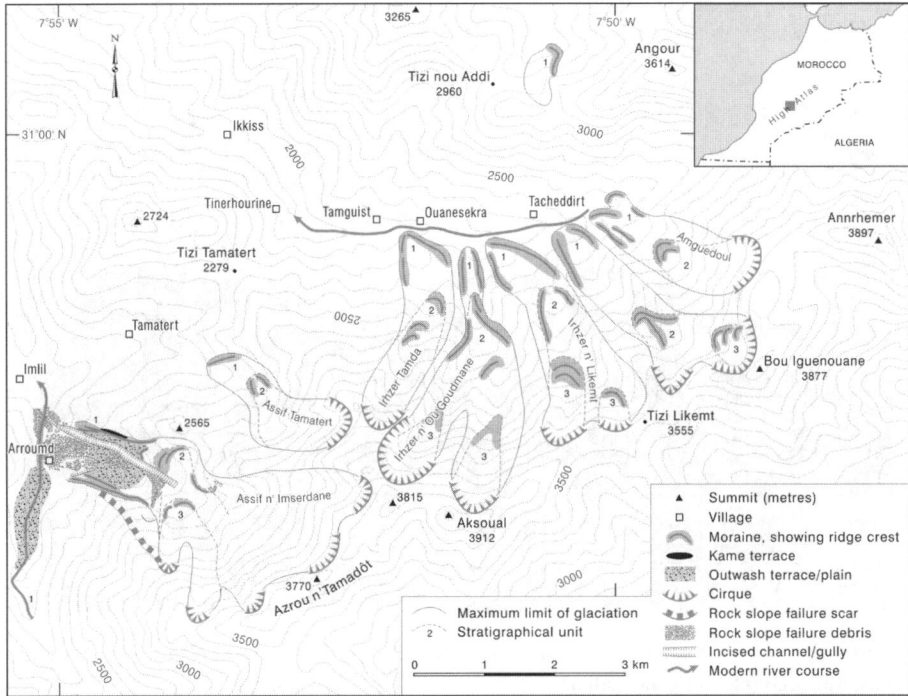

Figure 3.2 Moraines on the northern slopes of Aksoual (3912 m asl), High Atlas, Morocco. Cosmogenic nuclide exposure ages (^{10}Be) indicate that some of the youngest moraines (marked as 3 on the map) are Younger Dryas in age (12.9–11.7 ka). Reproduced from Hughes *et al.* (2011b), with permission

and oldest moraines have yielded ^{10}Be exposure ages much older than the LGM (Hughes *et al.*, 2011b).

3.2.2 Iberia

The former glaciers of Iberia are particularly important for understanding Pleistocene cold-stage climates. This is because of their proximity to the North Atlantic and the fact that the Polar Front, important in the development of depression weather systems, migrated south as far as western Iberia (Ruddiman and McIntyre, 1981) during cold periods. This has important implications for moisture supply to the Mediterranean region during Pleistocene cold-climate events, which have long been the subject of intense debate and speculation (e.g. Butzer, 1957; Tzedakis, 2007).

One of the largest Pleistocene ice caps in Iberia formed over the Sierra Segundera and Sierra de la Cabrera in the Sanabria National Park (Cowton *et al.*, 2009). There is evidence for at least three phases of glaciation in this area. The most

Figure 3.3 Cirque moraines on the northern slopes of Bou Iguenouane (3877 m asl) in the High Atlas, Morocco (photo by the author)

Figure 3.4 Landslide deposits resulting from a rockslope failure in a glaciated valley near Arroumd in the Toubkal Massif, High Atlas, Morocco (photo by the author)

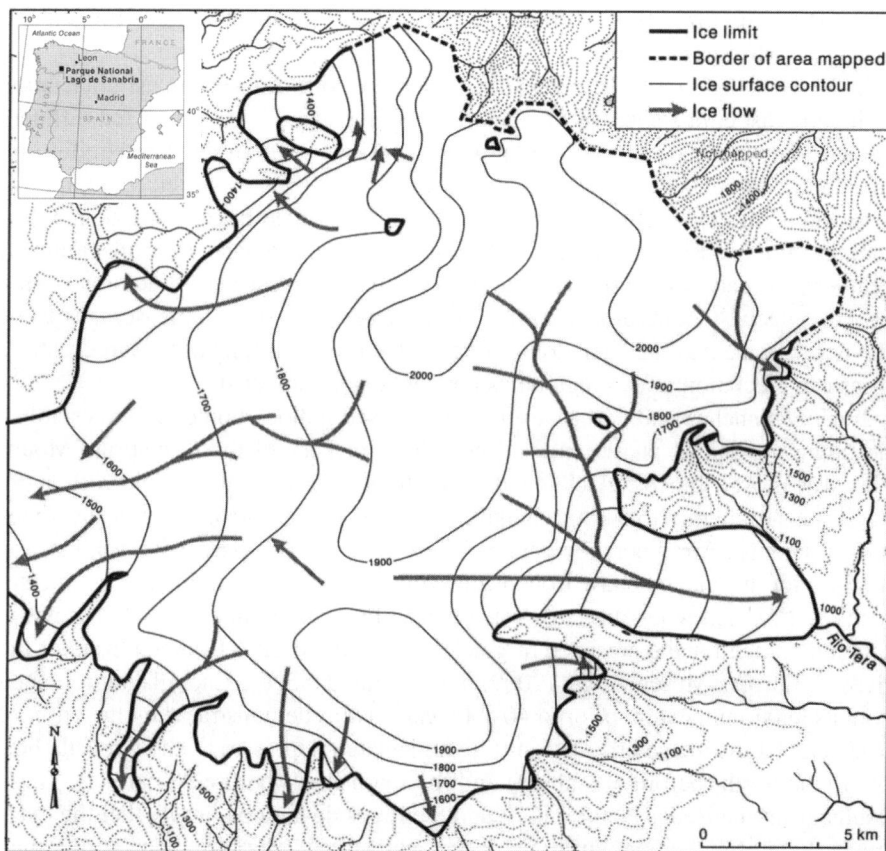

Figure 3.5 The Pleistocene Sanabria ice cap over the Sierra de la Cabrera and Sierra de Segundera in northwest Spain. Reproduced from Cowton *et al.* (2009), with permission from Elsevier

extensive glaciation was characterized by a large plateau ice cap, which covered an area of more than $440 \, \text{km}^2$, with a maximum ice thickness of c. 300 m and outlet glaciers reaching as low as 1000 m (Figure 3.5). Radiocarbon dates from the base of lacustrine sequences appear to suggest that the most extensive phase of ice-cap glaciation occurred during the last cold stage (Weichselian), with deglaciation occurring before 14–15 ka ^{14}C BP. A second phase of glaciation is recorded by the moraines of valley glaciers, which may have drained small plateau ice caps; whilst a final phase of glaciation is recorded by moraines in the highest cirques.

West of the Sanabria National Park are the Queixa (Galicia, Spain) and Gerês (Portugal) massifs. Here, [21]Ne and [10]Be cosmogenic isotope analyses have been applied to date glacially polished bedrock surfaces and push-moraine boulders. Three glacial phases have been identified, with the oldest two being of Middle Pleistocene age and the youngest glaciation dating from the last cold stage of the Pleistocene

(Vidal-Romaní *et al.*, 1999; Fernández-Mosquera *et al.*, 2000; Vidal-Romaní and Fernández-Mosquera 2006).

In the Serra da Estrela, the highest mountains of Portugal (1991 m asl), glaciation during the last cold stage was characterized by a plateau ice cap that fed diffluent glaciers. The ice cap covered a total area of 66 km^2 with an average equilibrium line altitude of 1650 m. The ice cap and its outlet glaciers have produced impressive erosional landforms including deep glacial troughs formed by ice more than 300 m thick in some places (Vieira, 2007). Thermoluminescence dates from glaciofluvial sediments of between 16.6 ± 2.5 and 10.6 ± 1.6 ka suggest glacial activity during the Late-glacial (Vieira *et al.*, 2001). These were the first radiometric ages to be published on the Estrela glaciations, although the precise timing of the largest phase of ice cap glaciation in this region has not yet been established.

Jiménez-Sanchez and Farias Arquer (2002) used radiocarbon dating to establish the timing of glacial phases in the Redes Natural Park of the Cantabrian Mountains. The most extensive glacial phase was characterized by an ice field with outlet glaciers extending up to 5 km in length, descending to c. 950 m asl with snowlines at c. 1550 m asl. A radiocarbon date of 29 ± 0.2 ka ^{14}C BP (34.2 ± 0.5 cal. ka) was obtained for this glacial phase from a core retrieved from ice-dammed lacustrine deposits that formed when drainage was blocked by a lateral moraine. This provides a minimum age for the presence of glacier ice. In the nearby Comella basin of the Picos de Europa, Moreno *et al.* (2009) demonstrated that the Enol glacier retreated from its maximum extent prior to 40 000 years ago as demonstrated by the onset of proglacial lacustrine sedimentation in two glaciated depressions: the Comella hollow to the north (before 40 cal. ka) and the Lago Enol (before 38 cal. ka). These results, from the Redes Natural Park and the Picos de Europa, imply that the maximum phase of glaciation during the last glacial stage took place much earlier than the global LGM. However, in other massifs of Spain such as the Sierra de Gredos and Sierra de Guadarrama, ^{36}Cl exposure ages from glacial landforms suggest that glaciers reached their maximum extent close to the global LGM (Palacios *et al.*, 2010, 2011). A similar dichotomy of geochronologies has also occurred in the Pyrenees (see below).

3.2.3 Pyrenees

The Pyrenees were extensively glaciated during the Pleistocene and some of the first glacial geomorphological research was undertaken in the nineteenth century by Penck (1885). Former glaciations have been attributed to at least two Pleistocene cold stages, the oldest deposits to the Rissian Stage and a second, higher suite of deposits to the Würmian Stage (Barrère, 1963). However, the deposits of the older glaciation are poorly preserved and the glaciers of the last cold stage were close in size to the most extensive glaciations. The largest Pleistocene glaciers formed on the northern slopes of the Pyrenees in France (Delmas *et al.*, 2011). For example, in the Ariège valley, glaciers extended 62 km to an altitude of c. 300–400 m asl (Hérail *et al.*, 1986; Delmas *et al.*, 2011).

Several studies from lake and bog sequences within former glacier limits on both the northern and southern slopes of the Pyrenees have produced radiocarbon ages suggesting maximum glacier advances that were asynchronous with the record of global ice volume – in several areas preceding the global LGM by tens of thousands of years (e.g. Mardones and Jalut, 1983; Vilaplana, 1983; Andrieu *et al.*, 1988; Jalut *et al.*, 1988; 1992; Montserrat 1992; Reille and Andrieu, 1995; García Ruiz *et al.*, 2003; González-Sampériz *et al.*, 2006). However, some authors have questioned this interpretation, such as Turner and Hannon (1988, p. 57), who suggested that the old ^{14}C ages, first reported in Mardones and Jalut (1983), are likely to have been affected by hard-water error or possibly a mixture of reworked organic matter – a view that was later reiterated by Pallàs *et al.* (2007) who applied cosmogenic nuclides to date glacial landforms.

Initially research suggested that cosmogenic nuclide analyses appeared to conflict with the dating frameworks that argue for an 'early' glacier maximum in the Pyrenees and suggest a local glacier maximum *after* 25 000 years ago and close in time to the global LGM at c. 21 ka. Pallàs *et al.* (2007) presented 25 ^{10}Be exposure analyses for granodioritic glacial erosion surfaces and boulders in the Upper Noguera Ribagorçana Valley in the south-central Pyrenees. On the lowermost glacial units, the oldest exposure age was 21 ± 4.4 ka and was obtained from an erratic boulder. Similar findings were reported by Delmas *et al.* (2008) from the Carlit massif in the eastern Pyrenees. However, more recent data by these researchers has revealed evidence for an earlier glacial maximum during the last glacial cycle (Pallàs *et al.*, 2010; Delmas *et al.*, 2011) and the initial contradictions between radiocarbon and cosmogenic ages appear to have been the result of insufficient data from comparable sites.

There is growing evidence for an 'early' glacier maximum in the Pyrenees. In the Cinca and Gállego River valleys (south-central Pyrenees and Ebro Basin, Spain), Lewis *et al.* (2009) interpreted optically stimulated luminescence (OSL) ages obtained from glacial deposits and loess as indicating glacial periods at 85 ± 5 ka, 64 ± 11 ka, 36 ± 3 ka and 20 ± 3 ka. They found that maximum extent of glaciers during the last glacial occurred at 64 ± 11 ka and this provides independent support for dating frameworks from other glacial sequences in the Pyrenees.

3.2.4 Maritime Alps

The Maritime Alps in southeast France and northwest Italy represent the southernmost extension of the European Alps. They are situated only 50 km from the coast at Monaco and are strongly influenced by Mediterranean climate systems. The highest peak is Argentera (3297 m asl) and this mountain supported some small glaciers as recently as the twentieth century (see Section 3.3 on modern glaciers, below).

During the last Pleistocene cold stage the Maritime Alps were extensively glaciated and the ice masses were contiguous with the main Alpine ice sheet. Lacustrine sediments at Lac Long Inférieur, a glaciated cirque (2090 m asl) in the French Maritime Alps, indicate that ice had melted by 14.19 ± 0.13 ka $_{14}$C BP (Ponel et al., 2001). Glacier retreat was punctuated by several readvances, and on the Italian side of the Argentera massif, Fisinger and Ribolini (2001) documented evidence for several glacial advances during the Late-glacial substage. In Italy, outlet glaciers from the Maritime Alps' ice sheet descended the Gesso and Stura di Demonte valleys and nearly to the Po Plain. The lowest frontal and lateral moraines in the Gesso Valley have recently been dated using ^{10}Be nuclide analysis (Granger et al., 2006). Boulders sampled on two separate moraine crests gave mean ^{10}Be cosmogenic ages of 16.3 ± 0.9 and 18.8 ± 1.0 ka. Granger et al. (2006) argued that these two moraines represent the inner and outer moraine crests of the same glacial event. The older maximum age of the Gesso glacier is broadly consistent with the Alps further north, where the northern lobe of the Rhône glacier reached its maximum extension between c. 21.0 and 19.1 ka around the time of the global LGM (Ivy-Ochs et al., 2004). Later glacier advances at higher elevations in the Gesso valley have also now been dated using ^{10}Be exposure age dating (Federici et al., 2008). Four boulder samples from the Piano del Praiet frontal moraine gave a weighted mean age of 11.3 ± 0.4 ka. Federici et al. (2008) suggest that the frontal moraine correlates with the Egesen Stadial of the Alps (Younger Dryas) and they argue that this suggests a synchronicity of the Egesen deglaciation across the European Alps.

3.2.5 Italian Apennines

Traces of glaciation are recorded throughout the Italian Apennines – extending from the Ligurian Alps in the north to the Appennino Calabro in the south. The Apennines reach an altitude of 2912 m asl at Corno Grande in the Gran Sasso, central Italy. Here, Giraudi and Frezzotti (1997) have demonstrated that the maximum glacier extent of the last cold stage (Würmian) occurred prior to 22.7 ± 0.6 ka ^{14}C BP. At the glacier maximum, ice in the Campo Imperatore area of the Gran Sasso extended 10.5 km down-valley and covered an area of 19 km^2 with an equilibrium line altitude (ELA) of c. 1750 m. A series of recessional moraines and rock glaciers in the Gran Sasso are thought to correspond to periods of glacier stabilization or readvance between 20 and 10 ka ^{14}C BP.

Middle Pleistocene glacial deposits older than those considered to be of Würmian age are usually less well preserved and are often strongly eroded, smoothed and smaller in size (Federici, 1980). Kotarba et al. (2001) presented U-series ages for calcite cements within these moraines as old as 135 ± 10 ka. Thus, given that the cements formed after the host deposits were formed, the moraines supported correlation with the Rissian Stage of the Alps (Kotarba et al., 2001). It is therefore clear that at least two major glacial advances are recorded in the Gran Sasso area, with readvance also recorded during the Würmian Late-glacial.

In the Campo Felice, in the central Apennines, the borehole data of Giraudi *et al.* (2010) suggest that five glacial advances respectively occurred during MIS 2, 3–4, 6, 10 and 14. This interesting dataset has shown that the glacial history of the Italian Apennines is more complex than previously thought.

3.2.6 Dinaric Alps

Large areas of the Balkans were glaciated during the Pleistocene. Some of the largest glaciers and ice caps formed in Montenegro, which is one of the most heavily glaciated countries of the Mediterranean region. Large Pleistocene glaciers formed in the Prokletije Mountains, on the border of Montenegro and Albania, where glaciers extended over 35 km, forming U-shaped valleys, moraines and numerous glacial lakes (Cvijić, 1913). However, Milivojević *et al.* (2008) have questioned the extent of ice cover suggested by Cvijić and argue that glaciers were more restricted in size. For example, Milivojević *et al.* argue that deposits bounding Lake Plav in the northern Prokletije, Montenegro, are remnants of alluvial fans and not moraines as previously thought.

Further west, on Mount Durmitor (2530 m asl), there is extensive evidence of glaciation, and 18 glacial lakes are present in this area (Djurović, 2009). An extensive ice cap covered much of this area as well as neighbouring massifs, such as Sinjajevina, Kapa Morača and Maganik to the south. In the latter two areas, the glacial limits on the western slopes were traced by Liedtke (1962a). This major ice cap glaciation occurred during the Middle Pleistocene (Hughes *et al.*, 2011a) and the contiguous ice caps of central Montenegro covered an area greater than 1400 km². Moraines of this major glaciation occur at altitudes between 500 and 1000 m. Late Pleistocene glaciation was significant but restricted to the higher valleys of the main massifs (Figure 3.6).

Near the Adriatic coast on Mount Orjen (1895 m asl), glacial cirques and valleys exploited older karstic forms to support extensive glaciers (Penck, 1900; Sawicki, 1911; Menkovic *et al.*, 2004; Hughes *et al.*, 2010). The extensive glaciation on Mount Orjen (1895 m asl) was probably a function of very high precipitation, and modern values here exceed 5000 mm (Magaš, 2002). The largest, and oldest, glaciation expanded over most of Orjen forming an ice cap that covered an area greater than 150 km². Nearby coastal mountains such as Lovćen (1749 m asl) were also glaciated during the Pleistocene (Liedtke, 1962b). Again, the most extensive phase of glaciations in these coastal mountains appears to have occurred during the Middle Pleistocene, whilst smaller valley and cirque glaciers formed during the last cold stage of the Pleistocene.

Marjanac and Marjanac (2004) presented evidence for glaciation in the coastal Dinaric Alps of Croatia. Significantly, Marjanac and Marjanac (2004) describe features on some of the Croatian coast and islands in the Adriatic, which they argue are glacial in origin. These include kame-terraces on the Krk and Pag islands to the west of the Velebit Mountains, as well as glacial and periglacial deposits on the

Figure 3.6 Late Pleistocene moraines in the Durmitor massif, Montenegro. These moraines formed during the Younger Dryas (12.9–11.7 ka) (photo by the author). (*A full colour version of this figure appears in the colour plate section*)

mainland coast nearby at Novigradsko More and Karinsko More. The glacial record in the Balkans is not yet fully understood, in terms of both extent and chronology. However, it is clear that this region supported some of the largest glaciers and ice caps in the Mediterranean mountains.

3.2.7 Bulgaria

Large areas of southern Bulgaria drain into the Mediterranean Sea and the mountains were extensively glaciated during the Pleistocene. Some of the first studies of glaciation in this area were made by Jovan Cvijić (1896, 1908, 1909) and later by Louis (1930). More recently, Velčev (1995) has suggested that multiple glaciations are recorded in Bulgaria including Middle Pleistocene glaciations (during the Mindel and Riss cold stages) as well as Late Pleistocene glaciation (Würmian stage). However, the timing of glaciations in Bulgaria has not yet been widely tested using geochronological techniques. The ELA of the last phase of major glaciation in the Rila Mountains was in the range 2100–2300 m asl, which is higher than in the western Mediterranean and the Adriatic and may be explained by drier conditions compared with these regions (Kuhlemann *et al.*, 2008).

3.2.8 Greece

In Greece, glacial deposits and landforms, from the northern Pindos to Crete, have been identified by numerous workers (see review by Woodward and Hughes, 2011, and references therein). Hughes *et al.* (2006b) correlated the various glacial and periglacial units recorded on Mount Tymphi in northern Greece with cold-stage intervals recorded in the pollen stratigraphy at nearby Ioannina (Table 3.1).

The oldest and most extensive glaciation in Greece is called the Skamnellian Stage, after the village of Skamnelli, and the type section (representative exposure of glacial deposits of this age) occurs in the deposits south of the village at c. 39°54′08″N, 20°50′40″E, 984 m asl. This glaciation has been correlated with Marine Isotope Stage (MIS) 12, which was a major period of glaciation in Europe between c. 480 and 430 ka. Ice covered an area of 59.3 km^2 and formed an ice field over the Tymphi plateau between the peaks of Gamila (2497 m asl) and Astraka (2436 m asl) from where ice tongues spilled down into the nearby gorges above the Aoos and Vikos canyons. The mean ELA of the glaciers on Mount Tymphi during the Skamnellian Stage was 1741 m asl. Hughes *et al.* (2007b) argued that the large glaciation of the Skamnellian Stage was forced by very low temperatures, with summer temperatures at least 11°C lower than modern summer values.

The next cold stage recorded by glacial deposits in Greece is termed the Vlasian Stage after the mountain of Vlasi (c. 2200 m asl) separating the Vourtapa and upper Laccorponti valleys on Mount Tymphi. The type locality for this stage is situated on the eastern side of Vlasi, at c. 39°55′50″N, 20°51′10″E, 1650 m asl in the Vourtapa valley, where moraines are clear and well preserved (Figure 3.7). This glaciation has been correlated with MIS 6, which occurred between c. 190 and 127 ka. Other glaciations may have occurred in the mountains of Greece between MIS 12 and MIS 6, but there is no record of these glaciations in the northern Pindos Mountains and thus any glaciers, if they formed, are likely to have been smaller than the Vlasian Stage glaciers. On Mount Tymphi, glaciers were significantly smaller during the Vlasian Stage compared with the Skamnellian Stage and covered a total area of 21.3 km^2, less than half the area of the Skamnellian Stage ice. They were just over 120 m higher than the latter glaciers, with an ELA of 1862 m asl.

The most recent glaciation on Mount Tymphi occurred during the Tymphian Stage, which is named after the mountain itself. Most of the glacial and periglacial deposits are situated in cirques close to the mountain's central and highest ridge. The type locality is characterized by the moraines and relict rock glaciers of the Tsouka Rossa cirque, at c. 39°58′45″N, 20°50′40″E, 2025 m asl. The glaciers were small during this cold stage and covered an area of just 3.9 km^2. This is because the ELA was positioned close to the highest ridges, at c. 2174 m asl. Eight rock glaciers also formed at the same time as these glaciers and had a lower latitude of c. 1800 m asl. Some of these rock glaciers developed from moraines formed by higher cirque glaciers whilst others formed at the base of steep scree slopes. Hughes *et al.* (2003, 2006c) used this evidence to reconstruct climate during the Tymphian Stage – the last ice age in Greece – and found that temperatures were 8.9°C lower than today

Table 3.1 Pleistocene cold-stage chronostratigraphy for Greece based on a combination of glacial events in the Pindos Mountains and a continuous record of environmental change at Ioannina in northwest Greece

Age (× 1,000 years)	MIS	Ioannina (IN 249/284)	Para-stratotype boundary (IN 249)	Pindos Chrono-stratigraphy	Local Stratotype
11.5–	1	Holocene	17.25 m		
73.9–11.5	2			**Tymphian Stage**	Tsouka Rossa Member 39°58′45″N, 20°50′40″E, 2025 m a.s.l.
	3				
	4				
73.9–83.0*	5a	Interstadial 2§			
88.5*–83.0*	5b	*Stadial 2§*			
104.5*–88.5*	5c	Interstadial 1§			
111.0*–104.5*	5d	*Stadial 1§*	45.88 m		
126.6*–111.0*	5e	Metsovon	59.00 m		
189.6–126.6*	6		76.00 m	**Vlasian Stage**	Vourtapa Member 39°55′50″N, 20°51′10″E, 1650 m a.s.l.
244.2–189.9	7a-e	IN-26 Zitsa IN-23a		?	
303–244.2	8	Katara			
339–303	9a-e	IN-17 Pamvotis			
362–339	10	Dodoni I/II			
423–362	11		162.75 m	**Skamnellian Stage**	Kato Radza Member 39°54′08″N, 20°50′40″E, 984 m a.s.l.
478–423	12		184.00 m		

Reproduced from Hughes et al. (2006c) with permission from Elsevier.

MIS, marine isotope stage.

§Denotes names based on Tzedakis et al. (2002) – all other names for the Ioannina sequence are from Tzedakis (1994).

*Denotes interval dates from Tzedakis et al. (2002) – all other dates from orbitally-tuned marine isotope records (Imbrie et al., 1984; Martinson et al., 1987).

Figure 3.7 Vlasian Stage (190–127 ka) moraines in the Vourtapa valley on Mount Tymphi in northwestern Greece (photo by the author). (*A full colour version of this figure appears in the colour plate section*)

whilst annual precipitation would have been similar to modern values at more than 2000 mm.

Hughes *et al*. (2006d) found evidence of a similar glacial sequence to Mount Tymphi on neighbouring Mount Smolikas (2637 m asl). Here, there is also evidence for a fourth and later glacial phase in the highest cirques, where small cirque glaciers developed with an ELA of c. 2372 m asl and total area of less than 0.5 km^2. This phase of glaciation is likely to have occurred after the glacial maximum of the Tymphian Stage, possibly during the Younger Dryas (12.9–11.7 ka) (Hughes *et al*., 2006d).

The glacial history of Mount Olympus (2917 m asl), the highest mountain in Greece, was investigated by Smith *et al*. (1997). Here, glaciers extended to the piedmont during the most extensive glacial phase and altitudes as low as 500 m asl. Smith *et al*. (1997) proposed a tentative chronology for glaciation on Mount Olympus by correlating soils on glacial deposits with dated soils in the river deposits of the Larissa Basin. The oldest and most extensive glaciation was correlated with soils older than 200 ka, leading Smith *et al*. (1997) to suggest that this glaciation occurred during MIS 8. A second phase of glaciation, characterized by upland ice and valley glaciers that did not reach the piedmont, was correlated with MIS 6. During the last major glacial phase, glaciers were restricted to valley heads, and this glacial phase was correlated with MIS 4 to 2. Smith *et al*. (1997) suggested

that a further set of moraines at c. 2200 m in the high cirque of Megali Kazania may
be Holocene Neoglacial features. However, features at a similar altitude on Mount
Smolikas (2632 m asl) in the Pindos Mountains were argued to be of Late-glacial
age (Hughes *et al.*, 2006d).

3.2.9 Turkey

The mountains of Turkey cover a vast area spanning from the Mediterranean Sea to
the borders of Georgia, Armenia, Iran and Iraq. Where the Mediterranean influence
ends and begins is difficult to establish, although even the mountains of far northeastern Turkey are likely to have been influenced, at least in part, by Mediterranean
climate systems during the Pleistocene.

In the Kaçkar Mountains, bordering the Black Sea (in NE Turkey Figure 3.1),
Akçar *et al.* (2007) have presented ^{10}Be ages from granitic surfaces in the Kavron
Valley on the southern slopes of the highest peak, Kaçkar Dağı (3932 m asl).
The Kavron Valley glacier advance began at least 26 ± 1.2 ka and continued until 18.3 ± 0.9 ka. After this time the glacier retreated and by 15.5 ± 0.7 ka the main
valley became ice-free with glaciers restricted to tributary cirques. A later glacier
advance took place between 13.0 ± 0.8 and 11.5 ± 0.8 ka. A similar situation has
been reported from a northward-draining valley of Mount Verçenik (3907 m asl), the
second highest peak in the Kaçkar Mountains. Here, a valley glacier advanced before 26.1 ± 1.2 ka and continued to advance until 18.8 ± 1.0 ka. According to Akçar
et al. (2008), the Verçenik glacier collapsed rapidly and after 17.7 ± 0.8 ka the main
valley was ice-free with only five small glaciers in tributary valleys. Among these,
the Hemşin glacier had completed its retreat by c. 15.7 ± 0.8 ka. This was followed
by a Late-glacial advance, identified on the basis of glacial erosion features (Akçar
et al., 2008).

In western Turkey, in the north-facing valleys of Mount Sandıras (2295 m asl),
Late Pleistocene glaciers extended to c. 1.5 km in length and terminated at an altitude of c. 1900 m asl. Sarıkaya *et al.* (2008) have utilized *in situ* cosmogenic ^{36}Cl
to ascertain the exposure age of boulders on moraine surfaces in these valleys and
found that the last local glacier maximum occurred at approximately 20.4 ± 1.3 ka.
In this valley, glacier retreat was interrupted by readvances at c. 19.6 ± 1.6 ka and
16.2 ± 0.5 ka. Glacier modelling suggests that temperatures during the last glacier
maximum on Mount Sandıras were 8.5–11.5°C lower than modern values and that
precipitation was nearly double that of today (Sarıkaya *et al.*, 2008).

On Mount Erciyes in central Turkey, Sarıkaya *et al.* (2009) found evidence of
four periods of glaciation. Glaciations on this extinct volcano were dated using
^{36}Cl nuclide analysis to provide exposure ages for 44 moraine boulders. The largest
glacier formed close in time to the LGM and reached 6 km in length and descended
to an altitude of 2150 m asl. These glaciers began to retreat 21.3 ± 0.9 ka. The
glaciers then readvanced and retreated by 14.6 ± 1.2 ka (Late-glacial), and again
by 9.3 ± 0.5 ka during the Early Holocene. The last glacier advance took place at

3.8 ± 0.4 ka, although a retreating glacier still survives in the northwestern cirque of the mountain.

3.2.10 Lebanon

Moraines exist in the high mountains of Lebanon in the eastern Mediterranean. Messerli (1966) reported the presence of moraines on both the highest peak in Lebanon, Qornet es Saouda (3088 m asl), in the Jbel Liban, and also on Mount Hermon (2814 m asl) in the south of the country. In both mountain areas, moraines are present between altitudes of 2500 and 2750 m. Glaciers would have been quite small on Mount Hermon, with lengths of less than 1 km. On Qornet es Saouda, however, glaciers reached several kilometres in length in the Rjoum valley on the northwestern flank of the mountain. Messerli (1966) tentatively suggested that two different glaciations were recorded in the Lebanon. Very little, if any, glacial research has been done in this area since the work of Messerli.

3.2.11 The glaciated Mediterranean islands

Corsica was extensively glaciated during the Pleistocene. Glacial features on the island were first reported by Pumpelly (1859), and the most recent published studies include those by Heybrock (1954), Letsch (1956) and Conchon (1986). Kuhlemann *et al.* (2005) used geomorphological evidence to reconstruct the former glaciers of Corsica. Large ice fields and valley glaciers formed during the Würmian glacial maximum, and some glaciers were up to 14 km long. The ELA of the Würmian glaciers was between 1400 and 1750 m asl, with variations attributed to precipitation differences. The Late Pleistocene age of the Corsican glacial landforms has been confirmed by Kuhlemann *et al.* (2008) using [10]Be exposure dating.

In Crete, the southernmost of the large Mediterranean islands, evidence of glaciation is limited and, in some areas, disputed. The clearest evidence of former glaciation is on Mount Idi (or Mount Psiloritis, 2456 m asl), the highest mountain in Crete. Here, Fabre and Maire (1983) recognized a cirque and associated moraines, the latter at an altitude of c. 1945 m asl. Further to the west, in the White Mountains, the evidence of glaciation is ambiguous, if present at all. Poser (1957), Bonnefont (1972) and Boenzi *et al.* (1982) did not find evidence of glaciation. However, Nemec and Postma (1993) have argued that the White Mountains were glaciated during the Pleistocene. They studied a series of alluvial fans and argued that they were formed by large water discharges associated with ice-cap melting (for more details see review in Hughes *et al.*, 2006a).

On other Mediterranean islands there is no reported evidence of former glaciation. Mount Etna (3329 m asl), on Sicily, is by far the highest mountain on the Mediterranean islands, and indeed one of the highest mountains of the Mediterranean region, yet no glacial evidence has been reported. However, this is not to say that glaciers did not form on this mountain during the Pleistocene (it is quite likely that glaciers did indeed form), only that the evidence, on what is an active volcano, has not been preserved.

3.3 Modern and recent glaciers in the Mediterranean mountains

Small glaciers, such as those in the Mediterranean mountains, respond rapidly to climate change and their behaviour provides important insight into climatic changes in this important climatic region (Hughes and Woodward, 2009; Hughes, 2011). The modern glaciers of this region are small and many have disappeared in the twentieth century. However, some do still survive and these are described below.

3.3.1 Iberia (Spain/France)

A glacier survived in the Sierra Nevada of southern Spain in the Corral Veleta cirque (37°N, 3°E) until the 1920s (see Figure 3.1). This was the southernmost glacier in Europe at that time. In the Picos de Europa, in northern Spain, four ice patches still survive (see Figure 3.1). However, the only 'true' remaining glaciers in Iberia are found in the Pyrenees (see Figure 3.1). Here, more than 20 glaciers covered a total area of 495 ha at the beginning of the twenty-first century (González Trueba et al., 2008). The Maladeta Massif, which includes the highest peak in the Pyrenees – Pico de Aneto (3404 m asl) supports some of the largest glaciers. However, in the Maladeta Massif, glaciers retreated dramatically through the twentieth century and glaciers shrunk by 530 ha between 1894 and 2001 (González Trueba et al., 2008).

3.3.2 Maritime Alps (France/Italy)

According to Federici and Pappalardo (1995) 15 small glaciers were present in the Maritime Alps on the border of France and Italy (see Figure 3.1). However, the modern state of these glaciers is unclear. These glaciers are the most southerly in the European Alps and some are situated less than 50 km from the Mediterranean coast. Most of the glaciers are found in the Argentera Massif, which contains the highest peaks of the Alpes Maritimes, including Cime de Argentera (3297 m asl). At the turn of the century, the ELA of the six largest Argentera glaciers was c. 2800 m asl (Fisinger and Ribolini, 2001). As with many other mountain glaciers of the Mediterranean, all of the Maritime Alps glaciers have retreated during the last century since the Little Ice Age. ELAs during the eighteenth and nineteenth centuries were 100–150 m lower than at the turn of the twenty-first century (Federici and Pappalardo, 1995; Pappalardo, 1999).

3.3.3 Apennines (Italy)

The southernmost glacier in Europe is often attributed to the Calderone glacier, which is situated in a north-facing cirque below the highest peak of the Italian Apennines, Corno Grande (2912 m) (see Figure 3.1). Like many other Mediterranean glaciers, the Calderone glacier retreated markedly through the twentieth century. Between 1916 and 1990 the volume of the glacier is estimated to have been reduced by about 90% and the area by about 68% (Gellatly *et al.*, 1994; D'Orefice *et al.*, 2000). According to Pecci *et al.* (2008), in the last decade, the glacier has split into two portions and the remaining ice bodies are rapidly shrinking (Pecci *et al.*, 2008).

3.3.4 Julian Alps (Slovenia)

In the Julian Alps of Slovenia, two small glaciers exist below the peaks of Triglav (2864 m asl) and Skuta (2532 m asl) (see Figure 3.1). On Triglav, the highest peak in Slovenia, the Zeleni Sneg glacier is situated on the northern slopes between c. 2550 and 2400 m asl. It retreated throughout much of the twentieth century (Sifrer, 1963). In 1995 the glacier covered an area of only 3.3 ha (Gabrovec, 1998). By 2007 the glacier area had reduced to only 0.6 ha. However, harsher than average winters in the past decade have slowed glacier retreat and in the last few years the glacier has actually expanded in area (Triglav-Čekada *et al.*, 2011). The Skuta glacier has exhibited similar retreat to the Triglav glacier during the twentieth century (Pavšek, 2004). However, given the recent very harsh winters in this region, the future state of these glaciers will provide interesting research.

3.3.5 Dinaric and Albanian Alps (Montenegro/Albania)

A small glacier exists in the Durmitor massif in northern Montenegro (Figure 3.8) (also see Figure 3.1). This glacier survives in a north-facing cirque below the mountain of Sljĕme (2455 m). In 2005 this glacier covered an area of c. 5.0 ha and dramatically reduced in size following a hot summer in 2007 (Hughes, 2008). However, in 2006 the glacier was only 25–50 m up-valley of moraines that formed during the Little Ice Age (Hughes, 2007) when at least eight glaciers were present in the Durmitor massif (Figure 3.9) (Hughes, 2010). In northern Albania, several small glaciers are present in the Prokletije Mountains (North Albanian Alps). At least four glaciers are present around the highest peak, Maja Jezercë (2694 m), the largest covering an area of c. 5.4 ha (Milivojević *et al.*, 2008; Hughes, 2009). The Montenegrin and Albanian glaciers survive at very low altitudes for this latitude. They are sustained by avalanching and windblown snow in addition to strong shading (Hughes, 2008, 2009). It is likely that more glaciers exist in these mountains – especially in the Prokletije – and further research is required.

Figure 3.8   The Debeli Namet glacier in the Durmitor massif, Montenegro. Photograph taken in September 2006 (photo by the author)

3.3.6   Pirin Mountains (Bulgaria)

Two small glaciers are present in the Pirin Mountains of Bulgaria along with numerous permanent snow patches (Grunewald *et al.*, 2004, Gachev *et al.*, 2010) (see Figure 3.1). The largest of these, the Sneschnika glacier, is found below the northern cliffs of Vihren (2914 m asl), the highest peak in the Pirin, and covered an area of just 1 ha in 2005. This glacier is currently the southernmost such feature in Europe. Another glacier is present below the peak of Kutelo (2908 m asl) just a few kilometres north of Vihren. The glaciers and perennial snow fields of this area are found in cirques at altitudes between 2400 and 2750 m asl.

3.3.7   Turkey

The largest concentration of glaciers in the Mediterranean mountains is found in Turkey, where at least 40 glaciers are present. The largest of these occur in the mountains of eastern Turkey (Akçar and Schlüchter, 2005) (see Figure 3.1). In the late twentieth century, satellite observations indicated that glaciers covered a total

Figure 3.9 Little Ice Age moraines below the northern cliffs of Zupci (2309 m asl) in the Durmi-tor massif, Montenegro. The Little Ice Age moraines are marked by the letter A. Older moraines of unknown age are indicated by the letter B. Reproduced from Hughes (2010), with permission

area of c. 22.9 km^2 in Turkey (Kurter, 1991). However, almost all glaciers have been in retreat since this survey (Çiner, 2004). On Mount Ararat, the highest peak in Turkey (5137 m asl), is a dormant volcano that supported an ice cap that covered c. 10 km^2 in the late twentieth century (Kurter, 1991). Glaciers are also found on other Turkish volcanoes, including Mount Süphan (4058 m asl) (where a glacier has formed in the crater) and Mount Erciyes (3917 m asl) (Kurter, 1991). However, the current state of these volcano glaciers is unclear. Some glaciers, such as one on Mount Erciyes, resemble rock glaciers (Kurter, 1991). Whether features such as these are simply debris-covered glaciers or true periglacial rock glaciers is not clear (cf. Hughes et al., 2003). Glaciers in Turkey are most numerous in the far southeast of the country near the borders with Iran and Iraq.

The largest glaciers in Turkey have formed on Mount Cilo (4135 m asl) where the Uludoruk and Mia Havara glaciers are nearly 4 km long and cover an area of 8 km^2. The Uludoruk glacier has retreated throughout the twentieth century and the altitude of the glacier front rose by c. 400 m between 1937 and 1991 (Çiner, 2004). Smaller glaciers also occur in the central Taurus, further to the west on the mountains of Aladag (3756 m asl) and Bolkardag (3524 m asl). The snowline in these areas is situated at c. 3450 m asl and glacier survival is strongly influenced by local topoclimatic controls (Kurter, 1991). In northeastern Turkey, at least 12 small glaciers exist in the Pontic Mountains, such as in the Kaçkar Massif (3932 m asl). According to Çiner (2004) glaciers in this massif cover a total area of about 2.54 km² with ELAs between 3100 and 3400 m.

3.4 Conclusion

The mountains of the Mediterranean have been shaped by multiple glaciations during the Pleistocene and some glaciers still survive today. Knowledge of glaciations has improved dramatically in recent years due to the development of new and existing dating techniques. However, there is still a lot to do and the glacial geochronology of many mountain areas remains unknown, or in some areas unresolved, with contradictions in the results of dating that has been applied. Glaciers have a very close relationship with climate – especially temperature and precipitation. Thus, the glacial record of the Mediterranean mountains has the potential to tell us a great deal about former atmospheric circulation in the region during Pleistocene cold stages. Future progress in this area of research will depend on the development of robust geochronologies and also detailed survey of former glacier extents.

As for the last remaining glaciers of the Mediterranean mountains, the future looks bleak. Future simulations of climate in the Mediterranean region suggest drier and hotter conditions for the period 2071–2100. In many areas, winter precipitation is expected to decrease whilst summer temperatures are also predicted to rise, with a rise in mean June–August temperatures of 5°C expected for some areas (Giorgi and Lionelli, 2008). If this scenario happens then the remaining glaciers will disappear and there will be no perennial ice left in the Mediterranean mountains. Whether this will happen remains to be seen, especially since the complexities of mountain topoclimates mean that the future behaviour of small mountain glaciers is often difficult to predict (Hughes, 2008, 2009).

References

References marked as bold are key references.

Akçar, N. and Schlüchter, C. (2005) Paleoglaciations in Anatolia: a schematic review and first results. *Eiszeitalter und Gegenwart* 55:102–121.

Akçar, N., Yavuz, V., Ivy-Ochs, S., Kubik, P.W., Vardar, M. and Schlüchter, C. (2007) Paleoglacial records from Kavron Valley, NE Turkey: Field and cosmogenic exposure dating evidence. *Quaternary International* 164-165:170–183.

Akçar, N., Yavuz, V., Ivy-Ochs, S., Kubik, P.W., Vardar, M. and Schlüchter, C. (2008) A case for a down wasting mountain glacier during Termination I, Verçenik Valley, NE Turkey. *Journal of Quaternary Science* 23:273–285.

Allen, J.R.M., Huntley, B. and Watts, W.A. (1996) The vegetation and climate of northwest Iberia over the last 14 000 yr. *Journal of Quaternary Science* 11:125–147.

Andrieu, V., Hubschman, J., Jalut, G. and Herail, G. (1988) Chronologie de la déglaciation des Pyrénées françaises. *Bulletin de l'Association Française pour l'étude du Quaternaire* 2-3:55–67.

Barrère, P. (1963) La période glaciaire dans l'ouest des Pyrénées centrales franco-espagnoles. *Bulletin de la Societe Géologique de France* 7:516–526.

Boenzi, F., Palmentola, G., Sanso, P. and Tromba, F. (1982) Aspetti geomorfologici del massiccio dei Leuka Ori nell'isola di Creta (Grecia), con particolare reguardo alle forme carsiche. *Geologia Applicata e Idrogeologia* 17:75–83.

Bonnefont, J.C. (1972) La Crète, etude morphologique. PhD thesis, Bibliothèque universitaire centrale, Université de Lille III, 845 pp.

Bordonnau, J. (1992) *Els complexos glacios-lacustres relacionats amb el darrer cicle glacial als pirineus.* Logroño: Geoforma Ediciones, 251 pp.

Butzer, K.W. (1957) Mediterranean pluvials and the general circulation of the Pleistocene. *Geografiska Annaler* 39:48–53.

Calvet, M. (2004) The Quaternary glaciation of the Pyrenees. In: Ehlers, J. and Gibbard, P.L. (eds), *Quaternary Glaciations – Extent and Chronology. Part I: Europe.* Amsterdam: Elsevier, pp. 119–128.

Celerier, J. and Charton, A. (1922) Sur la présence de formes glaciaires dans le Haut Atlas de Marrakech. *Hespéris* 2:373–384.

Celerier, J. and Charton, A. (1923) Un lac d'origine glaciaire dans le Haut Atlas (lac d'Ifni). *Hespéris* 3:501–513

Çiner, A. (2004) Turkish glaciers and glacial deposits. In: Ehlers, J. and Gibbard, P.L., (eds), *Quaternary Glaciations – Extent and Chronology. Part I: Europe.* Amsterdam: Elsevier, pp. 419–429.

Conchon, O. (1986) Quaternary glaciations in Corsica. *Quaternary Science Reviews* 5:429–432.

Cowton, T., Hughes, P.D. and Gibbard, P.L. (2009) Palaeoglaciation of Parque Natural Lago de Sanabria, northwest Spain. *Geomorphology* 108:282–291.

Cvijić, J. (1896) Tragovi starih gletcera na Rili. *Glas Srpske Kraljevske Akademije Nauka* 54, Belgrade.

Cvijić, J. (1908) Grundlinien der Geographie und Geologie von Mazedonien und Alt-serbien. *Petermanns Geographische Mitteilungen* 162:1–391.

Cvijić, J. (1909) Beobachtungen über die Eiszeit auf der Balkan-Halbinsel, in den Südkarpathen und auf dem mysischen Olymp. *Zeitschrift für Gletscherkunde* 3:1–35.

Cvijić, J. (1913) The ice age in the Prokletije and surrounding mountains (in Serbian). *Glas Srpske Kraljevske Akademije Nauka* XCI:1–149.

Delmas, M., Gunnell, Y., Braucher, R., Calvet, M. and Bourlès, D. (2008) Exposure age chronology of the last glaciation in the eastern Pyrenees. *Quaternary Research* 69:231–241.

Delmas, M., Calvet, M., Gunnell, M., Braucher, R. and Bourlès, D. (2011) Palaeogeography and [10]Be exposure-age chronology of Middle and Late Pleistocene glacier systems in the northern Pyrenees: implications for reconstructing regional palaeoclimates. *Palaeogeography, Palaeoclimatology, Palaeoecology* 305:109–122.

Djurović, P. (2009) Reconstruction of the Pleistocene glaciers of Mt. Durmitor in Montenegro. *Acta Geographica Slovenica* 49-2:263–289.

D'Orefice, M., Pecci, M., Smiraglia, C. and Ventura, R. (2000) Retreat of Mediterranean glaciers since the Little Ice Age: Case study of Ghiacciaio del calderone, Central Apennines, Italy. *Arctic, Antarctic and Alpine Research* 32:197–201.

Ehlers, J. (1996) *Quaternary and Glacial Geology.* Chichester: John Wiley & Sons, Ltd, 578 pp.

Fabre, G. and Maire, R. (1983) Néotectonique et morphologénèse insulaire en Grèce: le massif du Mont Ida (Crète). *Méditerranée* 2:39–40.

Federici, P.R. (1980) On the Riss glaciation of the Apennines. *Zeitschrift für Geomorphologie* 24:111–116.

Federici, P.R. and Pappalardo, M. (1995) L'evoluzione recente dei ghiacciai delle Alpi Maritime. *Geografia Fisica e Dinamica Quaternaria* 18:257–269.

Federici, P.R., Granger, D.E., Pappalardo, M., Risolini, A., Spagnolo, M. and Cyr, A.J. (2008) Exposure age dating and Equilibrium Line Altitude reconstruction of an Egesen moraine in the Maritime Alps, Italy. *Boreas* 37:245–253.

Fernández-Mosquera, D., Marti, K., Vidal Romani, J.R. and Weigel, A. (2000) Late Pleistocene deglaciation chronology in the NW of the Iberian Peninsula using cosmic-ray produced [21]Ne in quartz. *Nuclear Instruments and Methods in Physics Research* B172:832–837.

Fisinger, W. and Ribolini, A. (2001) Late glacial to Holocene deglaciation of the Colle Del Vei Bouc-Colle Del Sabbione Area Argentera massif, Maritime Alps, Italy – France. *Geografia Fisica e Dinamica Quaternaria* 24:141–156.

Gabrovec, M. (1998) The Triglav glacier between 1986 and 1998. *Geografski zbornik* 38:89–110.

Gachev, E., Gikov, A., Zlatinova, C. and Blagoev, B. (2011) Present state of Bulgarian glacierets. *Landform Analysis* 11:16–24.

García-Ruiz, J.M., Valero-Garcés, B.L., Martí-Bono, C. and González-Sampériz, P. (2003) Asynchroneity of maximum glacier advances in the central Spanish Pyrenees. *Journal of Quaternary Science* 18:61–72.

Gellatly, A.F., Smiraglia, C., Grove, J.M. and Latham, R. (1994) Recent variations of Ghiacciaio del Calderone, Abruzzi, Italy. *Journal of Glaciology* 40:486–490.

Giorgi, F. and Lionelli, P. (2008) Climate projections for the Mediterranean region. *Global and Planetary Change* 63:90–104.

Giraudi, C. and Frezzotti, M. (1997) Late Pleistocene glacial events in the Central Apennines, Italy. *Quaternary Research* 48:280–290.

Giraudi, C., Bodrato, G., Lucchi, M.R. *et al.* (2011) Middle and late Pleistocene glaciations in the Campo Felice Basin (central Apennines, Italy). *Quaternary Research* 75:219–230.

González-Sampériz, P., Valero-Garcés, B.L., Moreno, A. *et al.* (2006) Climate variability in the Spanish Pyrenees during the last 30,000 yr revealed by the El Portalet sequence. *Quaternary Research* 66:38–52.

González Trueba, J.J., Martín Moreno, R., Martínez de Pisón, E. and Serrano, E. (2008) Little Ice Age glaciation and current glaciers in the Iberian Peninsula. *The Holocene* 18:551–568.

Granger, D.E., Spagnolo, M., Federici, P., Pappalardo, M., Ribolini, A. and Cyr, A.J. (2006) Last glacial maximum dated by means of [10]Be in the Maritime Alps (Italy). *Eos Transactions. American Geophysical Union 87, Fall Meeting Supplement*, Abstract H53B-0634.

Grunewald, K., Weber, C., Scheithauer, J. and Haubold, F. (2006) Mikrogletscher im Piringebirge (Bulgarien). *Zeitschrift für Gletscherkunde und Glazialmorphologie* 39:99–114.

Hérail, G., Hubschman, J. and Jalut, G. (1986) Quaternary glaciation in the French Pyrenees. *Quaternary Science Reviews* 5:397–402.

Heybrock, W. (1954) Firnverhältnisse auf Korsika. *Zeitschrift für Gletscherkunde und Glazialgeologie* 3:75–78.

Hughes, P.D. (2007) Recent behaviour of the Debeli Namet glacier, Durmitor, Montenegro. *Earth Surface Processes and Landforms* 32:1593–1602.

Hughes, P.D. (2008) Response of a Montenegro glacier to extreme summer heatwaves in 2003 and 2007. *Geografiska Annaler* 192:259–267.

Hughes, P.D. (2009) 21ˢᵗ century glaciers and climate in the Prokletije Mountains, Albania. *Arctic, Antarctic and Alpine Research* 41:455–459.

Hughes, P.D. (2010) Little Ice Age glaciers in Balkans: low altitude glaciation enabled by cooler temperatures and local topoclimatic controls. *Earth Surface Processes and Landforms* 35:229–241.

Hughes, P.D. (2011) Mediterranean glaciers and glaciation. In: Singh, V.P., Singh, P. and Haritsaya, U.K. (eds), *Encyclopedia of Snow, Ice and Glaciers*. Berlin: Springer, pp. 726–730.

Hughes, P.D and Woodward, J.C. (2008) Timing of glaciation in the Mediterranean mountains during the last cold stage. *Journal of Quaternary Science* 23:575–588.

Hughes, P.D. and Woodward, J.C. (2009) Glacial and periglacial environments. In: Woodward, J.C. (ed.), *The Physical Geography of the Mediterranean*. Oxford: Oxford University Press, pp. 353–383 (Chapter 12).

Hughes, P.D., Gibbard, P.L. and Woodward, J.C. (2003) Relict rock glaciers as indicators of Mediterranean palaeoclimate during the Last Glacial Maximum (Late Würmian) of northwest Greece. *Journal of Quaternary Science* 18:431–440.

Hughes, P.D., Gibbard, P.L. and Woodward, J.C. (2005) A formal stratigraphical approach for Quaternary glacial records in mountain regions. *Episodes* 28:85–92.

Hughes, P.D., Woodward, J.C. and Gibbard, P.L. (2006a) Glacial history of the Mediterranean mountains. *Progress in Physical Geography* 30:334–364.

Hughes, P.D., Woodward, J.C., Gibbard, P.L., Macklin, M.G., Gilmour, M.A. and Smith, G.R. (2006b) The glacial history of the Pindus Mountains, Greece. *Journal of Geology* 114:413–434.

Hughes, P.D., Woodward, J.C. and Gibbard, P.L. (2006c) Late Pleistocene glaciers and climate in the Mediterranean region. *Global and Planetary Change* 46:83–98.

Hughes, P.D., Woodward J.C. and Gibbard, P.L. (2006d) The last glaciers of Greece. *Zeitschrift für Geomorphologie* 50:37–61.

Hughes, P.D., Gibbard, P.L. and Woodward, J.C. (2007a) Geological controls on Pleistocene glaciation and cirque form in Greece. *Geomorphology* 88:242–253.

Hughes, P.D., Woodward, J.C. and Gibbard, P.L. (2007b) Middle Pleistocene cold stage climates in the Mediterranean: new evidence from the glacial record. *Earth and Planetary Science Letters* 253:50–56.

Hughes, P.D., Woodward, J.C., van Calsteren, P.C., Thomas, L.E. and Adamson, K. (2010) Pleistocene ice caps on the coastal mountains of the Adriatic Sea: palaeoclimatic and wider palaeoenvironmental implications. *Quaternary Science Reviews* 29:3690–3708.

Hughes, P.D., Woodward, J.C., van Calsteren, P.C. and Thomas, L.E. (2011a) The glacial history of the Dinaric Alps, Montenegro. *Quaternary Science Reviews* 30:3393–3412.

Hughes, P.D., Fenton, C.R. and Gibbard, P.L. (2011b) Quaternary glaciations of the Atlas Mountains, North Africa. In: Ehlers, J., Gibbard, P.L. and Hughes, P.D. (eds), *Quaternary Glaciations – Extent and Chronology, Part IV – A Closer Look*. Amsterdam: Elsevier, pp. 1065–1074.

Imbrie, J., Hays, J.D., Martinson, D.G. *et al.* (1984) The orbital theory of Pleistocene climate: support from a revised chronology of the marine δ¹⁸O record. In: Berger, A., Imbrie, J., Hays, J., Kukla, G. and Saltzman, B. (eds), *Milankovitch and Climate*. Dordrecht: Reidel, pp. 269–305.

Ivy-Ochs, S., Schäfer, J., Kubik, P.W., Synal, H–A. and Schlüchter, C. (2004) Timing of deglaciation on the northern Alpine foreland (Switzerland). *Eclogae Geologicae Helvetiae* 97:47–55.

Jalut, G., Andrieu, V., Delibrias, G., Fontugne, M. and Pagès, P. (1988) Palaeoenvironment of the valley of Ossau (western French Pyrénées) during the last 27,000 years. *Pollen et Spores* 30:357–394.

Jalut, G., Montserrat, J., Fontunge, M., Delibrias, G., Vilaplana, J. and Juliá, R. (1992) Glacial to interglacial vegetation changes in the northern and southern Pyrenees: deglaciation, vegetation cover and chronology. *Quaternary Science Reviews* 11:449–480.

Janssen, C.R. and Woldringh, R.E. (1981) A preliminary radiocarbon dated pollen sequence from the Serra Da Estrela, Portugal. *Finisterra, Revista Portuguesa de Geografia* 16/32:299–309.

Jiménez Sanchez, M. and Farias Arquer, P. (2002) New radiometric and geomorphic evidences of a last glacial maximum older than 18 ka in SW European mountains: the example of Redes Natural Park (Cantabrian Mountains, NW Spain). *Geodinamica Acta* 15:93–101.

Kolfschoten, T., Gibbard, P.L. and Knudsen, K.-L. (2003) The Eemian Interglacial: a Global perspective. Introduction. *Global and Planetary Change* 3:147–149.

Kotarba, A., Hercman, H. and Dramis, F. (2001) On the age of Campo Imperatore glaciations, Gran Sasso Massif, Central Italy. *Geografia Fisica e Dinamica Quaternaria* 24:65–69.

Kuhlemann, J., Frisch, W., Székely, B., Dunkl, I., Danišík, M. and Krumei, I. (2005) Würmian maximum glaciation in Corsica. *Austrian Journal of Earth Sciences* 97:68–81.

Kuhlemann, J. Rohling, E.J., Krumrei, I., Kubik, P., Ivy-Ochs, S. and Kucera, M. (2008) Regional synthesis of Mediterranean atmospheric circulation during the last glacial maximum. *Science* 321:1338–1340.

Kuhlemann, J., Milivojević, M., Krumrei, I. and Kubik, P.W. (2009) Last glaciation of the Šara Range (Balkan peninsula): increasing dryness from the LGM to the Holocene. *Austrian Journal of Earth Sciences* 102:146–158.

Kurter, A. (1991) Glaciers of the Middle East and Africa – glaciers of Turkey. In: Williams, R.S. and Ferrigno, J.G. (eds), *Satellite Image Atlas of Glaciers of the World*. United States Geological Survey Professional Paper 1386-G-1, pp. 1–30.

Kurter, A. and Sungur, K. (1980) Present glaciation in Turkey. *International Association of Hydrological Sciences* 126:155–160.

Letsch, K. (1956) Firnverhältnisse auf Korsika. *Zeitschrift für Gletscherkunde und Glazialgeologie* 3:268.

Lewis, C.J., McDonald, E.V., Sancho, C., Luis Peña, J. and Rhodes, E.J. (2009) Climatic implications of correlated Upper Pleistocene glacial and fluvial deposits on the Cinca and Gállego Rivers (NE Spain) based on OSL dating and soil stratigraphy. *Global and Planetary Change* 67:141–152.

Liedtke, H. (1962a) Eisrand und Karstpoljen am Westrand Lukavica-Hochfläche. *Erdkunde* XVI:289–298.

Liedtke, H. (1962b) Vergletscherungsspuren und periglazialerscheinungen am Südhang des Lovčen östlich von Kotor. *Eiszeitalter und Gegenwart* 13:15–18.

Louis, H. (1926) Glazialmorphologishche Beobachtungen im albanischen Epirus. *Zeitschrift der Gesellschaft für Erdkunde* 9/10:398–409.

Louis, H. (1930) Morphologische Studien in Südwest Bulgarien. *Geographische Abhandlungen, Stuttgart* 3:H. 2.

Magaš, D. (2002) Natural-geographic characteristics of the Boka Kotorska area as the basis of development. *Geoadria* 7/1:51–81.

Mardones, M. and Jalut, G. (1983) La tourbiére de Biscaye (alt. 409 m, Hautes Pyrénées): Approche paléoécologique des 45.000 dernières années. *Pollen et Spores* 25:163–212.

Marjanac, L. and Marjanac, T. (2004) Glacial history of the Croatian Adriatic and Coastal Dinarides. In: Ehlers, J. and Gibbard, P.L. (eds), *Quaternary Glaciations – Extent and Chronology. Part I: Europe*. Amsterdam: Elsevier, pp. 19–26.

Martinson, D.G., Pisias, N.G., Hays, J.D., Imbrie, J., Moore, T.C. and Shackleton, N.J. (1987) Age dating and the orbital theory of the ice ages: development of a high resolution 0–300,000 year chronostratigraphy. *Quaternary Research* 27:1–29.

Martonne, E. de (1924) Les formes glaciaires sur le versant Nord du Haut Atlas. *Annales de Géographie* 183:296–302.

Maull, O. (1921) Beiträge zur Morphologie des Peloponness und des Südlichen Mittelgriechenlands. *Geographische Abhandlungen* 10:119–129.

Menkovic, L., Markovic, M., Cupkovic, T., Pavlovic, R., Trivic, B. and Banjac, N. (2004) Glacial morphology of Serbia Yugoslavia, with comments on the Pleistocene glaciation of Montenegro, Macedonia and Albania. In: Ehlers, J. and Gibbard, P.L. (eds), *Quaternary Glaciations – Extent and Chronology. Part I: Europe.* Amsterdam: Elsevier, pp. 379–384.

Messerli, B. (1966) Das Problem der eiszeitlichen Vergletscherung am Libanon und Hermon. *Zeitschrift für Geomorphologie* 10:37–68.

Messerli, B. (1967) Die eiszeitliche und die gegenwartige Vertgletscherung im Mittelemeeraum. *Geographica Helvetica* 22:105–228.

Milivojević, M., Menković, L. and Ćalić, J. (2008) Pleistocene glacial relief of the central part of Mt. Prokletije (Albanian Alps). *Quaternary International* 190:112–122.

Monserrat, J.M. (1992) Evolución glaciar y postglaciar del clima y la vegetación en la vertiente sur del Pirineo: estudio palinológico. Monografías del Instituto Pirenaico de Ecologia 6. Consejo Superior de invetigaiones Cientíaragoza, Spain, 147 pp.

Moreno, A., Valero-Garcés, B.L., Jiménez-Sánchez, M. *et al.* (2010) The last deglaciation in the Picos de Europa National Park (Cantabrian Mountains, northern Spain). *Journal of Quaternary Science* 25:1076–1091.

Nemec, W. and Postma, G. (1993) Quaternary alluvial fans in southwestern Crete: sedimentation processes and geomorphic evolution. *Special Publication of the International Association of Sedimentology* 17:256–276.

Nowack, E. (1929) Die diluvialen Vergletscherungsspuren in Albanien. *Zeitschrift für Gletscherkund* 17:122–167.

Palacios, D., Marcos, J. de and Vázquez-Selem, L. (2010) Last Glacial Maximum and deglaciation of Sierra de Gredos, central Iberian Peninsula. *Quaternary International* 233:16–26.

Palacios, D., Andrés, N. de, Marcos, J. de and Vázquez-Selem, L. (2012) Glacial landforms and their palaeoclimatic significance inj Sierra de Guadarrama, Central Iberian Peninsula. *Geomorphology* 139–140:67–78.

Pallàs, R., Rodés, Á., Braucher, R. *et al.* (2007) Late Pleistocene and Holocene glaciation in the Pyrenees: a critical review and new evidence from [10]Be exposure ages, south-central Pyrenees. *Quaternary Science Reviews* 25:2937–2963.

Pallàs, R., Rodés, Á., Braucher, R. *et al.* (2010) Small, isolated glacial catchments as priority targets for cosmogenic surface exposure dating of Pleistocene climate fluctuations, southeastern Pyrenees. *Geology* 38:891–894.

Palmentola, G., Boenzi, F., Mastronuzzi, G. and Tromba, F. (1990) Osservazioni sulle tracce glaciali del M. Timfi, catena del Pindo (Grecia). *Geografia Fisica e Dinamica Quaternaria* 13:165–170.

Pappalardo, M. (1999) Remarks on the present-day condition of the glaciers in the Italian Maritime Alps. *Geografia Fisica e Dinamica Quaternaria* 22:79–82.

Pavšek, M. (2004) The Skuta glacier. *Geografski obzornik* 51:11–17.

Pecci, M., Agata, C. and Smiraglia, C. (2008) Ghiacciaio del Calderone (Apennines, Italy): the mass balance of a shrinking Mediterranean glacier. *Geografia Fisica e Dinamica Quaternaria* 31:55–62.

Penck, A. (1885) La Période glaciaire dans les Pyrénées. *Bulletin de la Societe d'Histoire Naturelle de Toulouse* 19:105–200.

Penck, A. (1900) Die Eiszeit auf der Balkanhalbinsel. *Globus* 78:133–178.

Ponel, P., Andrieu-Ponel, V., Parchoux, F., Juhasz, I. and De Beaulieu J.-L. (2001) Late-glacial and Holocene high-altitude environmental changes in Vallee des Merveilles Alpes-Maritimes, France: insect evidence. *Journal of Quaternary Science* 16:795–812.

Poser, J. (1957) Klimamorphologische Probleme auf Kreta. *Zeitschrift für Geomorphologie* 2:113–142.

Pumpelly, R. (1859) Sur quelques glaciers dans l' île de Corse. *Bulletin de la Société Géologique de France* 17:78.

Reille, M. and Andrieu, V. (1995) The late Pleistocene and Holocene in the Lourdes Basin, Western Pyrénées, France: new pollen analytical and chronological data. *Vegetation History and Archaeobotany* 4:1–21.

Ruddiman, W.F. and McIntyre, A. (1981) The North Atlantic during the last deglaciation. *Paleogeography, Paleoclimatology, Paleoecology* 35:145–214.

Sarıkaya, M.A., Zreda, M., Çiner, A. and Zweck, C. (2008) Cold and wet Last Glacial Maximum on Mount Sandıras, SW Turkey, inferred from cosmogenic dating and glacier modelling. *Quaternary Science Reviews* 27:769–780.

Sarıkaya, M.A., Zreda, M. and Çiner, A. (2009) Glaciations and paleoclimate of Mount Erciyes, central Turkey, since the Last Glacial Maximum, inferred from ^{36}Cl dating and glacier modeling. *Quaternary Science Reviews* 23–24:2326–2341.

Sawicki, R. von (1911) Die eiszeitliche Vergletscherung des Orjen in Süddalmatien. *Zeitschrift für Gletscherkunde* 5:339–350.

Sestini, A. (1933) Tracce glaciali sul Pindo epirota. *Bollettino della Reale Società Geografica Italiano* 10:136–156.

Sifrer, M. (1963) New findings about the glaciation of Triglav. *Geografiski zbornik* 8:157–210.

Smith, G.W., Nance, R.D. and Genes, A.N. (1997) Quaternary glacial history of Mount Olympus. *Geological Society of America Bulletin* 109:809–824.

Turner, C. and Hannon, G.E. (1988) Vegetational evidence for late Quaternary climatic changes in southwest Europe in relation to the influence of the North Atlantic Ocean. *Philosophical Transactions of the Royal Society, London* B318:451–485.

Tzedakis, P.C. (1994) Vegetation change through glacial-interglacial cycles: a long pollen sequence perspective. *Philosophical Transactions of the Royal Society of London* B345: 403–432.

Tzedakis, P.C. (2007) Seven ambiguities in the Mediterranean palaeoenvironmental narrative. *Quaternary Science Reviews* 26:2042–2066.

Tzedakis, P.C., Lawson, I.T., Frogley M.R., Hewitt, G.M. and Preece, R.C. (2002) Buffered tree population changes in a Quaternary refugium: evolutionary implications. *Science* 297:2044–2047.

Velčev, A. (1995) The Pleistocene glaciations in the Bulgarian mountains. *Annals University of Sofia, Faculty of Geology and Geography* 87:53–65.

Vidal Romaní, J.R., Fernández, D., Marti, K. and De Brum, A. (1999) Nuevos datos para la cronología glaciar pleistocena en el NW de la Península Ibérica. *Cadernos do Laboratorio Xeolóxico de Laxe* 24:7–29.

Vidal-Romaní, J.R. and Fernández-Mosquera, D. (2006) Glaciarismo Pleistoceno en el NW de la peninsula Ibérica (Galicia, España-Norte de Portugal). *Enseñanza de las Ciencias de la Tierra* 13:270–277; available at: http://www.raco.cat/index.php/ect/article/view/89058/133836.

Vieira, G.T. (2007) Combined numerical and geomorphological reconstruction of the Serra da Estrela plateau icefield, Portugal. *Geomorphology* 97:190–207.

Vieira, G., Ferreira, A.B., Mycielska-Dowgiallo, E., Woronko, B. and Olszak, I. (2001) Thermoluminescence dating of fluvioglacial sediments, Serra da Estrela, Portugal. *V REQUI – I CQPLI*, Lisboa, Portugal, 23–27 de Julho 2001, pp. 85–92.

Vilaplana, J.M. (1983) Quaternary glacial geology of Alta Ribagorçana Basin (Central Pyrenees). *Acta Geológica Hispánica* 18:217–233.

Walker, M., Johnsen, S., Rasmussen, S.O. *et al.* (2009) Formal definition and dating of the GSSP (Global Stratotype Section and Point) for the base of the Holocene using the Greenland NGRIP ice core, and selected auxiliary records. *Journal of Quaternary Science* 24:3–17.

Woodward, J.C. and Hughes, P.D. (2011) Glaciation in Greece: a new record of cold stage environments in the Mediterranean. In: Ehlers, J., Gibbard, P.L. and Hughes, P.D. (eds), *Quaternary Glaciations – Extent and Chronology, Part IV – A Closer Look*. Amsterdam: Elsevier: pp. 175–198.

Woodward, J.C., Hamlin, R.H.B., Macklin, M.G., Hughes, P.D. and Lewin, J. (2008) Glacial activity and catchment dynamics in northwest Greece: Long-term river behaviour and the slackwater sediment record for the last glacial to interglacial transition. *Geomorphology* 101:44–67.

4

Landforms and soils

Maria Teresa Melis and Stefano Loddo

4.1 Introduction

The landscape can be considered as the result of a competition between 'constructional' processes, such as tectonics, that act to increase or decrease the elevation of the crust, and 'destructional' processes, such as erosion, that modify it. These two processes interact and are influenced by other factors such as lithology, climate and sea-level changes. In a broader sense, mountains form a major component of terrestrial dynamism. Large mountain ranges as much as isolated peaks are highly exposed to environmental risk and degradation, mainly due to severe climate conditions and gravitational denudation processes. This chapter provides a description of the physical aspects of the Mediterranean mountain landscapes. The aim is to give an overview of their regional geomorphological dynamics, geological setting and pedological characteristics. In addition the chapter discusses some of the most typical mountain environments, ranging from the west to the east of the basin, with the inclusion of the massifs situated in two of the largest Mediterranean islands, Sicily and Sardinia.

4.2 Geomorphological processes in the Mediterranean mountain region

The morphology of the Mediterranean mountains is closely conditioned by the geodynamic thrusts still active and is easily readable by looking at the distribution of mountain ranges that surround this basin. The Pyrenees, the Alps, the Dinaric Alps, and the Hellenids, Atlas and Taurus mountains have evolved and have accompanied formation of the Mediterranean Basin. Their position provides a natural protection from the cold northerly perturbations of Europe thereby supporting the presence of mild climatic conditions. The highest peaks are found in the Alps, which border the Italian peninsula from west to east coming to lap the Italian Ligurian coast.

Mediterranean Mountain Environments, First Edition. Edited by Ioannis N. Vogiatzakis.
© 2012 John Wiley & Sons, Ltd. Published 2012 by John Wiley & Sons, Ltd.

The chains of the Pyrenees to the west and the Dinaric Alps and Hellenids to the east are the natural setting that overlooks the Mediterranean Basin, which includes the territories that, due to culture and geographical location, display similarities with western Asia and North Africa. Though belonging from the same orogenic events, the geomorphological evolution of Mediterranean mountain ranges was also strongly conditioned by the climate and its recent changes in the Quaternary (see Chapter 2).

It is customary to distinguish between the structural morphology, which depends on the curvature of the Earth's crust, and the glacial morphology and karst related to landslides. The first trend is manifested by the almost straight valleys and ridge lines, due to profound changes (faults and tectonic lines), while the layers of overlapping thrust belt do not usually present a clear morphological correspondence. The glacial morphology is demonstrated by the U-shaped valleys, with cirques, amphitheatres and glacial dorsal moraines (see Chapter 3). Lithology plays an important role in shaping mountain landscapes. Crystalline rocks (granite, gneiss, etc.) are resistant to erosion and are associated with prominent structures. Soft sedimentary schists and flysch have high erodibility and are prone to landslides, but produce good-quality soils. Resistant limestones, in contrast, result in soaring cliffs. On mountain foothills there are sandstone formations, resulting from the accumulation of material eroded by rivers and glaciers, which are mostly associated with soft landscapes.

4.3   Geological setting

Mountains dominate much of the Mediterranean area and are visible to travellers on land and sea alike. The Mediterranean Basin was shaped by the ancient collision between the northward-moving African-Arabian continent and the stable Eurasian continent. As Africa-Arabia moved north, it closed the former Palaeo-Tethys Ocean, which formerly separated Eurasia from the ancient supercontinent of Gondwana, of which Africa was a part. During the Permian along the eastern margin of Gondwana began the opening of the Neo-Tethys Ocean and the subduction of the Palaeo-Tethys Ocean. The collision of continents pushed up a vast system of mountains, extending from the Pyrenees in Spain to the Zagros Mountains in Iran. This episode of mountain building, known as the Alpine Orogeny, occurred mostly during the Oligocene (34 to 23 mya) and Miocene (23 to 5.3 mya) epochs. The Neotethys became larger during these collisions and associated folding and subduction. Tectonics and climate are the primary external processes governing mountain landscape evolution. Tectonic uplift creates elevated terrain and provides increased potential energy to the agents of erosion, such as fluvial systems (Quigley *et al.*, 2007), which are the source of most of the sediments and waters shaping and feeding lowland and coastal areas.

During the Quaternary period five major Pleistocene ice ages have occurred: Donau, Gunz, Mindel, Riss and Würm, named according to the moraine of the

Danube and some of its tributaries. Most of the traces we know are still on the most recent, namely Würm, which occurred from 75 000 to about 15 000 years ago. Of course, every major glacial advance corresponds to a very prolonged period (tens of thousands of years) and is characterized by a cold climate. The last significant advance occurred between 1590 and 1850, known as the Little Ice Age.

In a way glaciation has rendered the mountain environments more accessible to human activity, through widening and flattening of valleys, modelling large and accessible passes, excavation of features such as terraces along the slopes, and fertilization of otherwise barren land with moraine material.

4.4 Soil characteristics of Mediterranean landscape units

Soil formation is a function of climate, organisms, parent material, relief and time (Jenny, 1941, 1980), all of which determine soil properties by governing the type and intensity of the pedological processes involved. Since these soil-forming factors also govern geomorphic processes, landscape evolution is intimately related to soil development (McFadden and Knuepfer, 1990). Although all environmental factors may remain constant, pedological processes, operating over time, inevitably affect the soil-profile development at a given site.

There are some specific factors that characterize the evolution of Mediterranean soils, including clay illuviation and carbonate redistribution, in part of aeolian origin (Yaalon, 1997). Desert dust, incorporated within the soil profile, is one of the characteristic features of soil development. Erosion and truncation accompany severe deforestation especially on sloping lands, resulting in widespread shallow lithosols (Inceptisols and Entisols; Leptisols and Cambisols). Retransport of much of the eroded coarse-to-fine-grained colluvium and alluvium to footslopes and lowland results in the accumulation of materials serving the development of new, fertile soils suitable for agriculture. Where fine material accumulates, dark-coloured Vertisols or Xerolls, rather than red soils (Xeralfs), dominate the landscape. Several detailed studies have shown that the landscapes and soils of most Mediterranean regions are old, frequently polygenetic and affected by climatic fluctuation and also by humans' management of the landscape (Yaalon, 1997). A brief synthesis of pedological characterizations of mountain landscapes is given below: the soils are related to lithological units (parent material) for a simpler description. The USDA (United States Department of Agriculture) Soil Taxonomy classification is used (USDA, 1999).

4.4.1 Landscapes on limestone, dolomite and dolomitic limestone of the Palaeozoic and Mesozoic

These landscapes are generally rugged and characterized by harsh forms that alternate with gentle slopes and planes. Rocky outcrops occupy most of these landscapes and soils are limited to small areas exposed to the north where the forest is better conserved. The soils are predominantly of clay texture, with a lumpy and multi-faceted subangular and angular aggregation, and a good content of organic matter in the upper horizons where the natural vegetation is present, providing a useful depth of moisture and good fertility. The balance between soil, vegetation and erosion in these areas is very delicate. Any alteration due, for example, to fire, cutting or over-grazing, results in an intense and often irreversible degradation, leading to deserti-fication phenomena. Land use in these areas is mainly forestry while pastoral activ-ities are limited. Agriculture can only be practised on the gentler slopes and larger colluvial areas. These landscapes are mainly found in the eastern Mediterranean Basin in the mountain massifs of Greece and Lebanon, but also in western Sardinia.

4.4.2   Landscapes of the Palaeozoic metamorphic rocks

The landscapes that have developed on metamorphic rocks are characterized by a remarkable variety of forms reflecting differences in lithological nature and the in-fluence of morphological changes. The vegetation presents a number of different forms, depending on exposure, altitude, microclimate and anthropogenic activity. Generally we find less developed soils (Lithic, Typic and Dystric Xerorthents) on more siliceous rock types because of their greater resistance to alteration. The tex-tures vary from sandy to medium clay; the organic matter content is high only in the presence of bush or scrub forest. On sandstone and shale sandstone, landforms are gentler and the soils show greater development, especially in areas most heavily conserved. We find Typic and Lithic Xerochrepts where the slopes are less steep or there is more vegetation cover.

On these formations debris slopes are quite extensive and soils are generally highly evolved (Palexeralfs), with a relatively high erodibility due to the weak ag-gregation and low content of organic matter. The Pyrenees, Baetic chain and the At-las mountains are all characterized by outcrops on Palaeozoic metamorphic rocks.

4.4.3   Landscapes of intrusive rocks (granite, granodiorite, etc.)

Intrusive rocks are more homogeneous but the landscape is almost always charac-terized by steep slopes, deep valleys and rugged topography. Where the granite is altered it forms slope deposits, or cones, the landscape becomes gentler and less steep so that the land may be used in places for extensive or intensive agriculture. In this case, the most common soils in areas with steep slopes are Xerorthents (lithic and typic subgroups). More evolved soils lie beneath forest and generally belong to Typic or Dystric Xerochrepts. The degradation of these landscapes through erosion is particularly intense because of the fires, forestry operations and pasture improve-ment. These landscapes are present in both Sardinian and Corsican mountains.

4.4.4 Landscapes of effusive rocks

These landscapes are characterized by different rock types, with different forms and degrees of acidity. The landscapes are thus characterized by lava flows of basalt and acid forms with steep slopes on the more acid rocks and more or less harsh forms for the andesites. On flat surfaces of the highlands with basic rocks, soils belong mainly to the Lithic Xerorthents and to the Typic and Lithic Xerochrepts, alternated with long stretches of rock outcropping. Since these are soils with high fertility, they have always been used for grazing and partly for agriculture. This explains the presence of large areas with very thin soils and rocky outcrops, which are the result of widespread erosion. The soils on basalt or tuff on higher elevations may show andic characteristics due to the presence of amorphous materials (high water retention capacity, low density, high cation-exchange capacity and retention capacity for phosphorus, etc.). The landscape on andesites is characterized by a predominance of rocky outcrops, while at the base of mountains it is not uncommon to find Vertisols supporting an intensive agriculture. The ignimbrites have mixed forms: rough areas and sub-flat. In any case, rarely will you encounter evolved soils, but only forms of degradation. The natural vegetation, where it exists, generally comprises cork oak forests and more or less evolved scrub. These landscapes are mainly present in the mountains of southern Italy, where volcanism is active.

4.4.5 Landscapes on marly, arenaceous and calcareous-marly rocks

The landscape which has evolved over these rocks presents undulating characters in the marly-arenaceous formations and rugged forms on limestone formations. Sequences of soils are typical on the basis of morphology (chains) with Entisols at the top of the hills, Inceptisols on slopes and Vertisols in lowland areas. On the arenaceous rocks the chains are composed of Lithic Xerorthents at the top of the mountains, Calcic Xerochrepts on the slopes and Typic Xerochrepts at the base. Erosion is very active even though these substrates (mostly marls and sandstones) allow a quick change with the formation of soils in a relatively short time. Typical examples of these landscapes are found in the Italian Apennines where extreme erosional features such as badlands are present (e.g. 'Calanchi' and 'Bianchane').

4.4.6 Landscapes on detritus and debris cones

The quality of the soil depends on the lithological characteristics of the materials from which it originated – detritus, morphometric characteristics and vegetation. The soil has a rocky or stony surface. It may appear excessively drained and typically has a medium texture. The organic matter content in surface horizons is very high; they are classified as Mollisols; the high content of organic matter reduces the volume of soil penetrated by plant roots.

4.4.7 Landscapes on moraines

On the moraines at the foot of the mountains soils are common but variable in character. In the most ancient and evolved cases, the Alfisols are stable, sometimes desaturated in depth, with processes of acidification and leaching causing the evolution of Ultisols. In the youngest moraines or where there is ongoing erosion, the soils have reached a degree of evolution (Alfisols and Inceptisols). A good example is located in Serra Ivrea in northern Italy. Dating back to the Quaternary this is the largest glacial moraine of its kind in Europe.

4.5 Northern Mediterranean mountain landscapes: from the Pyrenees to the Hellenids

4.5.1 Pyrenees

The Pyrenees are divided into three longitudinal sections: the Western or the Atlantic Pyrenees, from the Bay of Biscay to the Pic d'Anie; the Central Pyrenees, from the Pic d'Anie to Col du Puymorens; and the Eastern or Catalan Pyrenees, from the Col du Puymorens to the Gulf of Lion in the Mediterranean sea. The Central Pyrenees include the highest summits of the chain: Pic de Néthou (3404 m) in the crest of Maladeta, Mont Posets (3375 m), and Monte Perdido, or Mont Voyager (3355 m).

Approximately 30–40 million years ago the same earth movements that created the Alps pushed up the Hercynian mountain mass submerged and covered by Secondary Era deposits. During this mountain-building era, the comparatively pliable sedimentary beds folded, whereas the much older and more rigid platform broke up, with hot springs bursting through the fracture lines.

The Axial Zone of the Pyrenean range is characterized by sharply defined ridges and bold peaks where a number of granitic extrusions have thrust through the more easily eroded primary sediments. However, limestone and schists of a particularly resistant type are also found among some of the highest summits; Monte Perdido is a good example, being Western Europe's highest limestone mountain. Nearby, the Cirque de Gavarnie (also limestone) reveals huge platforms of horizontal strata stacked one upon the other. The Pic du Midi d'Ossau owes its distinctive shape to an extrusion of volcanic rock. Glaciers of the Quaternary Ice Age (10 000 years ago) did much to give the range some of its most dramatic and appealing features. Although this glacial cover has almost completely disappeared today, there remain countless hanging valleys, more than 1000 tarns, cirques, sharp peaks and ragged crests that reveal the work of these ancient ice sheets. No less than the glaciers, water has helped shape the Pyrenees, and in many of the limestone massifs it is rainwater and snowmelt that have eroded the rock into a network of cracks and fissures, some of which penetrate to great depths and produce subterranean galleries and caverns, as those of Pierre St Martin. Other fissures become ravines and gorges, like those of the Basque country, or impressive canyons such as Añisclo and Escuain.

4.5.2 The Apennines

The Apennines extend down the entire Italian peninsula and are connected to the northwestern Alps, forming a pronounced curve with the concavity facing the Tyrrhenian Sea, and continuing beyond the Strait of Messina in the mountains of northern Sicily. The chain is approximately 1200 km long, 30 to 150 km wide and covers an area of some 150 000 km². The slopes are asymmetric: those of the Tyrrhenian side are usually longer and cut by longitudinal grooves interspersed with tracks from the upper reaches of tributaries draining into the Tyrrhenian Sea; the Adriatic side slopes down to the sea through a hilly region. Despite the morphological differences, the Apennine landscape finds its unity in the almost entirely sedimentary nature of the underlying rocks. Landforms created by water erosion become particularly evident in the clayey soils, where the valley flanks are cut by the furrows of the gullies and by landslides. When the soil is saturated by water content, movements of mass downstream are possible. The physical boundary between the Apennines and the Alps is conventionally regarded as the hill of Cadibona (436 m) near Savona. The geological boundary is marked by the Sestri-Voltaggio tectonic line. The Apennines orogenic cycle is determined by the Alpine orogeny. The expansion related to the formation of the Tyrrhenian basin has affected the entire margin of the Tyrrhenian Apennines, creating large faults, Horst-type structures (tectonic block up) and Graben (rift), and causing the rise of large amounts of magma that gave rise to the complex magmatic provinces of Tuscany, Lazio and Campania.

The Northern Apennines are composed mostly of sandstone and marly rocks that are easily eroded and have rounded shapes; there are abundant landslide soils. The Central Apennines, forming the main backbone, consist mainly of limestone. This different lithology and the general uplift confer specific traits to this part of the Apennines: in particular a more pronounced roughness of the landscape and the development of karst landforms. The highest peak of the Central Apennines is the Gran Sasso (2912 m at the summit of Big Horn), marking the culmination of the entire system and the seat of the only glacier in the Apennines.

The Southern Apennines splits into isolated forms, separated by deep basins without regular alignment. The geological structure is not uniform: although there are some limestone reliefs, as in the Central Apennines, there are also ancient crystalline rocks that are quite different from those making up the rest of the system. The Pollino massif in the Appennino Lucano, which rises to 2267 m in the Serra Dolcedorme, marks the southern end. Profoundly different from the rest of the

Apennines, in terms of both lithology and orography, are the Calabro Apennines, which consist essentially of crystalline rocks. Aspromonte (Montalto Mountain, 1955 m) is characterized by a domed top and steep slopes that descend to the sea. It forms the southern tip of the Italian peninsula. Finally, beyond the Strait of Messina, the Apennines continue along the northern coast of Sicily as the Siculus Apennines. They consist of the Peloritan Mountains, which are similar in geological structure to the Calabrian Apennines, the Nebrodi (or Caronie) and the Madonie mountains. These last are composed of limestone, marl and sandstone.

4.5.3 The Dinarids and the Hellenids

In the Balkan Peninsula the main mountain ranges are the Dinaric Alps (Outer and Inner Dinarids, Pelagonides and Hellenides), the Rila-Rodopi Massif, the Stara Planima Mountains and the East Balkan Uplands (Demek *et al.*, 1984). The young-fold mountains of the Dinarids run from northwest to southeast and about 21% of the land area lies above the treeline. Mesozoic limestone and dolomites are the main parent rock types; the Outer Dinarids have the largest contiguous area of karst in Europe. The Pelagonides run from Šar Planima (elevation 2748 m), a mountain range in the Balkans that extends from Kosovo and the northwest of the Republic of Macedonia to northeastern Albania, through Korab, its highest peak (2764 m), to Tajget (2409 m) in the south. The Hellenids form the backbone of mainland Greece, running north-northwest to south-southeast. South of the Peloponnese they disappear below sea level and take a sharp turn east, reappearing in the south Aegean island arc, conspicuously as the three large massifs of Crete; further east they continue in the Cilician and Taurus mountain ranges of southern Anatolia. In north-central Greece, two shorter, interrupted mountain ranges run more or less parallel to the Hellenids, and in the northeast are differently oriented ranges with a different geological history. The highest mountain is Olympus (2917 m), and there are numerous massifs and peaks above 2000 m throughout mainland Greece and Crete.

The most common soil types in the alpine and subalpine regions of the Balkan Mountains are: lithosols (widespread in the mountains of the Dinaric and Shara-Pindos system); Regosols on igneous rocks, crystalline schists and crystalline dolomites; rendzina on flysch and dolomites; ranker (humus-silicate soil) on siliceous rocks; calcomelanosol (limestone-dolomite black soil) on limestone and dolomite; podsol on acid igneous rocks; and brown earth on pure limestone and dolomite (Filipovski, 1996).

4.6 Eastern Mediterranean mountains: the Lebanon Mountains

The larger part of Lebanese territory is sloping, often steeply, with gradients ranging between 20 and 60%. However, the coastal area and inner Bekaa plain are relatively flat, with slope gradients less than 8%. From a morphostructural point of view, Lebanon consists of three units, two of them constituting the uplifted steep mountain ranges (Mount Lebanon and Anti-Lebanon), separated by the Bekaa depression. The western chain (Mount Lebanon) borders the Mediterranean Sea, displaying relatively gentle slopes on its western flanks and steeper ones on the eastern side. The highest point in Lebanon is in the northern part of this mountain chain, namely the Qornet es Saouda, 3083 m above sea level (asl). Precipitation (rain and snow) falls in abundance on the Lebanese mountain chains, especially on Mount Lebanon. The precipitation rate varies between 700 and 1200 mm/year with increasing elevations across Mount Lebanon; about 80% of the annual precipitation falls between November and February (Edgell, 1997).

Carbonate rocks (limestone and dolostone) dominate the known Lebanese rock succession. The oldest exposed rocks are of Liassic age (Dubertret, 1975). The Jurassic strata constitute the cores of the Mount Lebanon and Anti-Lebanon ranges. Cretaceous strata – especially the Cenomanian-Turonian Sannine and Maameltain formations – form the flanks of the mountain chains, covering most of the country's surface area. The stratigraphy of Lebanon has been investigated by a number of authors (Dubertret, 1955; Saint-Marc, 1974, 1980; Beydoun, 1988; Walley, 1997; Nader, 2000).

The Late Jurassic is characterized by a period of regional uplift, emergence and erosion. Upon emergence, the Jurassic rock mass was deeply fractured and karstified before volcanism took place. Subsequently, volcanic deposits and continental debris filled up the pre-existing fractures and depressions accentuating the palaeotopography (Renouard, 1955; Nader *et al.*, 2003).

The morphology of Mount Lebanon has been mainly affected by karstification, part of the meteoric diagenesis process, since the Miocene (Dubertret, 1975; Walley, 2001).

The Lebanese mountain range is one of the highest in the Mediterranean region, peaking at more than 3000 m, and the steepness of its slopes, combined with the rainfall it receives (up to 1500 mm/year), make these mountain lands an exceptional agro-ecosystem. The orography creates a mild climate, which differs from neighbouring arid and semiarid countries. Hence derive the pedoclimatic features of Lebanese soils. The soil temperature regime is mainly hyperthermic and the soil moisture regime, because of the typical Mediterranean seasonal rainfall distribution, is considered to be generally xeric. In the Mount Lebanon range the large areas of very shallow soils and rocky outcrops lead to episodes of severe erosion and land degradation. Recent studies have demonstrated that more than 50% of Lebanese territory is prone to desertification.

Considering the geological, morphological and climatic features of Lebanon, the country's pedological panorama can be synthetically described as follows. Soils classified as Petric Calcisols are mostly present in the northeastern Bekaa. Eutric Fluvisols, Vertic Cambisols and Vertisols appear especially in the central Bekaa. Eutric Luvisols, Rendzic Leptosols and Lithic Leptosols are the most widespread soils in the Lebanese mountains. Calcaric Cambisols, Eutric Vertisols, Calcaric Arenosols and Eutric Fluvisols characterize the soils of the coastal plain.

4.7   Southern Mediterranean mountains: Atlas Mountains and Iberian Baetic Cordillera

4.7.1   The Atlas

The Atlas mountain system in northwest Africa spans Morocco, Algeria and Tunisia. It is limited to the south by the Sahara desert, and to the north and west by the Mediterranean sea and Atlantic Ocean. Stretching for approximately 2500 km from southwestern Morocco to northern Tunisia, it forms a topographic and climatic barrier between the Mediterranean and the Sahara. It is part of the great Alpine-Himalayan folding: on the one hand, through the Rif, the Atlas can be considered as an extension of the Baetic system in the Iberian peninsula (Figure 4.1), while in its eastern section it is connected, under the Sicilian Channel, to the Italian Apennine system. In the Atlas traces of Precambrian and Caledonian bending are recognizable. The Hercynian orogeny later gave rise to the formation of a grand chain directed from north to south; its existence was short-lived, however, as it was already eroded and peneplanated by the end of the Palaeozoic. During the Mesozoic and Cenozoic, thick layers of sandstone, limestone and schist were deposited on top of this base, and then were folded during the Alpine orogeny. The first phase of this orogeny (Pyrenean phase) is of Eocene age and the hills were quickly eroded. During the Miocene the main orogenic phase occurred, which gave the Atlas its current configuration.

The structure of the Atlas system is quite complex. On its eastern side, there are two main mountain ranges, the northern coastal Atlas Tell, and the southern interior Saharan Atlas. The first extends parallel to the coast of Algeria, from the Gulf of Annaba to the border of Morocco. It is characterized structurally by overlapping

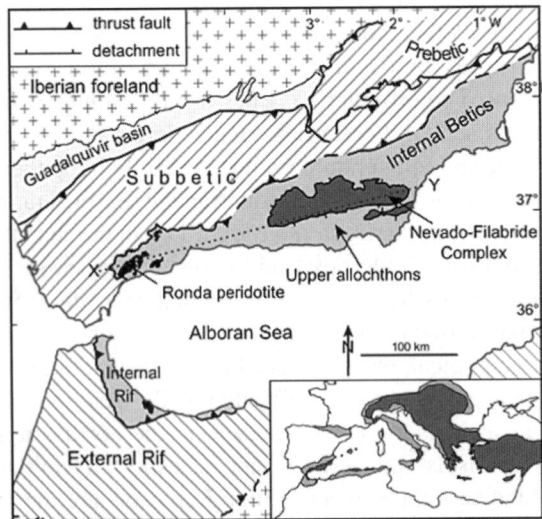

Figure 4.1   Tectonic map of Baetic-Rif arc. Reproduced from Platt *et al.* (2006), with permission

flaps and folds, including two massive arrays, separated by a longitudinal groove (the Cheliff depression); other main features are: the massif of Dahra, near Miliana; the Djurdjura (reaching 2308 m at Lalla Khadîdja, the highest point of Atlas Tell) and the mountains of Tlemcen (elevation 1841 m); Saida (1180 m); the massive Ouarsenis (1983 m); the Titeri mountains and the chain of Biban. The Atlas chain continues in Tunisia, in the mountains of Medjerda.

The Saharan Atlas mountains, separated from the Atlas Tell by a series of plateaus, stretch for approximately 700 km from the Moroccan border region to Biskra, beyond which they continue in the Aurès mountains. They consist of a regular succession of folds grouped in clumps, and reach a height of 2326 m at Mount Chélia. In the Western Atlas the Tell is split into a coastal range (Rif) and the Middle Atlas further inland. Between the two mountain chains, the corridor of Taza creates a natural route of communication between Morocco and Algeria. To the west the Saharan Atlas extends as the High Atlas, in a WSW–ENE direction, for approximately 750 km entirely in Morocco, from the plain of Tamlelt to Thira Cape. This is the highest part of the Atlas chain, with an average elevation of 2000 m and culminating at 4165 m s.m. in Mount Toubkal; the eastern sector is mainly composed of Jurassic limestone, while the western one is formed of crystalline rocks, raised a considerable height by powerful vertical movements. The presence of glacial cirques and valleys testifies to events of the glacial Neozoic.

The Atlas system is located in the path of oceanic air masses and receives abundant rainfall, especially on slopes exposed to the west. Its highest peaks are very snowy winter, and in the High Atlas are permanently covered with snow. The still active tectonics and the evolution of the climate during the Pleistocene are the main factors affecting the geomorphological dynamics of the landscapes. This is clearly evident in the forms shaped by the erosion of the slopes at lower altitudes, characterized by the presence of vertical walls engraved deep into the high mountain plateaus. This combination of morphologies is accompanied by the geographic location: a zone of transition between the temperate Mediterranean climate to an arid one, more properly related to the physiographic system of the Mediterranean coasts of Africa.

4.7.2 The Baetic Cordillera and the Sierra Nevada

The Baetic mountain chain is an impressive complex of ridges in the south and southeast of the Iberian Peninsula, extending approximately 1000 km from the Strait of Gibraltar to Cape Nao. It is bounded to the south by the Mediterranean, and to the north by the valley of the Guadalquivir River. It comprises several mountainous areas, mostly arid and barren, and includes the Sierra Nevada, which culminates in Mount Mulhacén at 3482 m.

The geological setting is linked to the tectonic collision during the Alpine Orogeny with the African plate. Since the Tortonian, the situation in the Baetic Cordillera and northern Morocco has been compressive in a NNW–SSE to N–S direction, combined with an east–west extension.

Figure 4.2 Pico de la Veleta (photo by the authors)

The Sierra Nevada, in the southern Mediterranean Iberian Peninsula, lies along the 37th parallel nearly east–west for about 78 km and occupies an area of 2000 km². The structural origin is highlighted in an asymmetry between the two large systems of northern and southern slopes. Characteristic morphostructural exposure led to the differentiation between the slopes that face the mainland interior and those that slope sharply towards the Mediterranean Sea. The Sierra Nevada mountain region is the third highest in Europe, reaching a maximum altitude at Mulhacén, but the peak best known for its distinctive shape is the Veleta (elevation 3394 m) (Figure 4.2). This massif hosted the last glacier in the Sierra Nevada, the Corral del Veleta, which was the southernmost glacier in Europe until its disappearance towards the end of the twentieth century. Various studies have been conducted to describe this small enclave in southeast Spain. Several geophysical, geothermal and geomorphological prospecting techniques were used to locate the permafrost at Corral del Veleta, an area highly sensitive to slight climatic variations, and to study its evolution under marginal conditions. The results confirm the location of Europe's southernmost permafrost remnant. The glacial cirque of the deceased Corral retains only areas of fossil ice and permafrost. In the warmer months, the snow disappears even from the top, revealing the lack of vegetation and rugged rocks.

The gradual retreat of the ice during the last glacial stage 10 000 years ago has revealed the different morphologies of overexcavation in the form of depressions giving rise to glacial lagoons that can be observed today (Figure 4.3). Glaciation in the Sierra Nevada was generally low intensity and the cirques grow very close to

Figure 4.3   A small glacial laguna with permanent water during the summer season (photo by the authors). (*A full colour version of this figure appears in the colour plate section*)

the ridge line. Consequently, the lagoons are at high altitudes, between 2600 and 3100 m. They are not very deep but there are many basins, and melting of snow and ice results in sheets of water that fill all the little depressions. During the summer the loss of water through seepage plus the action of the sun and wind causes many of these pools to dry out. The remaining ponds are restricted to a very small area (less than 100 km^2) around the highest peaks of the chain. The highest pond is Laguna del Corral at 3086 m, while at lower altitudes, between 2600 m and 2700 m, other ponds are located, including Lavanderos de la Reina, the Carnero, Misterioso and Puesto del Cura.

In the Sierra Nevada's Lanjarón Valley, Mulhacén Valley, Laguna Seca and Corral del Veleta, preglacial surfaces have been reworked to a greater or lesser degree and deposits formed during different cold episodes (Sanchez *et al.*, 1988; Sanchez, 1990; Gomez *et al.*, 1992). The soil parent materials are mainly composed of mica schists of the Nevado-Filabride Complex (Diaz de Federico and Puga, 1976). The Sierra Nevada summits, as described above, have an east–west alignment over a distance of approximately 90 km with an asymmetrical setting that divides the northern slopes from the southern ones. This morphology affects the climate of the different slopes; the dry conditions in the northern areas, exposed to the arid climate of central Andalusia, differ from those of the Mediterranean coastal areas. This environment, with a Mediterranean rainfall regime, mountain slopes and undercutting ephemerally flowing rivers, is a template for many of the mountain environments of the Mediterranean region, and results drawn from here could have widespread implications for many of the slopes in the region. The Alpujarras is a deep valley running parallel to and south of the crest of the Sierra Nevada (Figure 4.4). It is

Figure 4.4   The southern landforms of Sierra Nevada (photo by the authors)

drained by the deeply incised Guadalfeo river and its northern tributaries, and to the south is separated from the Mediterranean by the Sierra Contraviesa.

The mountain front shows well-developed triangular facets produced by stream dissection along the southwest border of the Sierra Nevada (Birot, 1965; Riley and Moore, 1993). These facets suggest active faulting during the Quaternary (Silva, 1994). Also suggestive are triangular facets at different elevations that suggest several stages of uplift and reactivation of tectonic process (Riley and Moore, 1993). The classic landscape related to the morphological phenomena of soil degradation is well recognizable on these slopes. Only in the alluvial terraces along the watercourses are some areas of soil present; in these areas an intensive agriculture has developed.

4.8   Characteristic mountain landscapes of the Mediterranean islands

4.8.1   Sardinian mountain landscape

The mountain landscape of Sardinia is characterized by the presence of the Gennargentu massif, a Palaeozoic massif whose peaks reach the highest heights at Bruncu Spina (1829 m) and Punta La Marmora (1834 m). The Gennargentu gives rise to the main waterways of eastern Sardinia: the Cedrino and the Flumendosa rivers.

Figure 4.5 The Mesozoic limestone and dolomite of Monte Corrasi (1463 m) dominate the village of Oliena, on the right side. The landscape on the plain is dominated by olive groves and pastures. On the mountain slopes afforestation with coniferous trees is dominant while on the hillsides some areas covered by oaks (*Quercus ilex*) are recognizable (photo by the authors)

Among the most beautiful areas is the Supramonte (Figure 4.5). It is an immense and wild chain comprising a dolomite limestone plateau that rises up to 1463 m. Soils of karst areas of the Supramonte are among the most highly vulnerable to degradation in Sardinia, where their regeneration, and the regeneration of vegetation, is obstructed by the presence of outcropping rock. Soil genesis in calcareous bedrock is an extremely slow process, with limestone solubilization occurring mainly in brief wet periods during the year.

It is possible to make some assumptions regarding the development patterns of soil profiles in relation to soil cover, taking into account the equally important role of substrata:

- On the upper slopes, where the shrub coverage is discontinuous, soil profiles are frequently truncated and soils have formed in small pockets within the rock fractures and are hence weakly developed (Lithic Entisols or Alfisols very leached, often Terra Rossa).

- At midslope, and more limitedly at the foot of the slopes, where plant cover is mainly composed of holm oak woods or macchia shrubland (maquis), the soil profile is generally better developed and more differentiated into horizons, for the most part argillic (Alfisols).

- At the foot of the slopes, where the original plant cover has been destroyed to make way for agropastoral activities, the soils are not well developed, and Inceptisols prevail.

- At the bottom of the valleys (e.g. the Lanaittu Valley), with the absence of plant cover, soils are often immature (Entisols). In some places they show deep profiles, which have allowed them to be used for agriculture.

In Sardinia the landscape morphology and climatic features make the island's soils very fragile and sensitive to uses that do not take into account the soil's suitability and limitations. In mountain areas the proper, balanced management of vegetation cover ensures the permanence of soil in situ, limiting the intensity of erosional phenomena. Due to these environmental conditions the degradation of soils is a problem of great concern on the island. Synthetically the Sardinian landforms have been divided in this way: 18.5% mountainous areas, 67.9% hilly areas and 13.6% plain areas (Pracchi and Terrosu Asole, 1971). Almost everywhere the soil moisture regime is of a xeric type (Raimondi *et al.,* 1995). Plain areas apart, in the remaining hilly and mountainous territory (86.4% of the total) soils belong mainly to the Cambisols, Leptosols and Regosols. These areas are considered marginal for most agricultural practices but have great importance for particular niche crops, and for grazing and forestry.

4.8.2 Active volcanic landscapes

In Mediterranean Basin two main volcanic regions can be identified: the Aegean and Italian. The Aegean region is characterized by reduced volcanic activity; it was the scene of violent eruptions in the past, notably the one that nearly destroyed the island of Santorini around 1400 BC, producing a large caldera. The Italian volcanic region, which runs from Tuscany to Sicily and Pantelleria, is the largest volcanic region in Europe and activity began here in the Pliocene (Woodward, 2009). Volcanic activity is still taking place, and we can distinguish three different types of volcanism.

The first type is the explosive volcanism of the Aeolian Islands, represented by two volcanoes, Stromboli and Vulcano, with different activities. The activity of Stromboli is continuous, with weak explosions sometimes followed by intermittent volcanic emissions. The activity of Vulcano develops instead into two stages: in the first stage, due to its viscosity, the lava forms a dome in its crater of stagnation, while in the second stage gas pressure underneath the dome causes it to explode and shatter, allowing the next release of lava. Mount Etna in eastern Sicily is the highest active volcano in Europe, and manifests the second type of volcanism, called effusive volcanism. The third kind of volcanic activity is the explosive volcanism of the Southern Tyrrhenian associated with Campi Flegrei, Ischia and, above all, Vesuvius. Eruptions some 35 000 years ago gave rise to pyroclasts, which accumulated and gave rise to the grey clay (Campanian ignimbrite), while subsequent eruptions have produced the yellow tuff that forms the base rock for the city of Naples. At Ischia volcanism produced the green tuff.

4.8.3   The Etna volcano: activity, soils and landscape

Mount Etna, located in eastern Sicily, reaches a height of about 3300 m asl. The active summit craters are closed westward of Bove Valley, a structure of collapse depth 1000 m, 7 km long and 5 km wide. The eruptions occur from the summit craters (Northeast Crater, Voragine, Bocca Nuova, Southeast Crater) and along the flanks of the volcano. Since the late 1970s, summit eruptions have increased in intensity, with extremely violent paroxysmal eruptive episodes after 1995, culminating in 2005 with 120 events, many of which have affected the southeast of the crater (Behncke and Neri, 2003; Branca and Del Carlo, 2004) (Figure 4.6).

These episodes have generated lava fountains accompanied by abundant pyroclastic flows transported a few tens of kilometres from the volcano (Figure 4.7). In addition to these intense manifestations, typically lasting less than 30 minutes, the Etna eruptive activity has affected both sides of the southwestern (Bove Valley) and

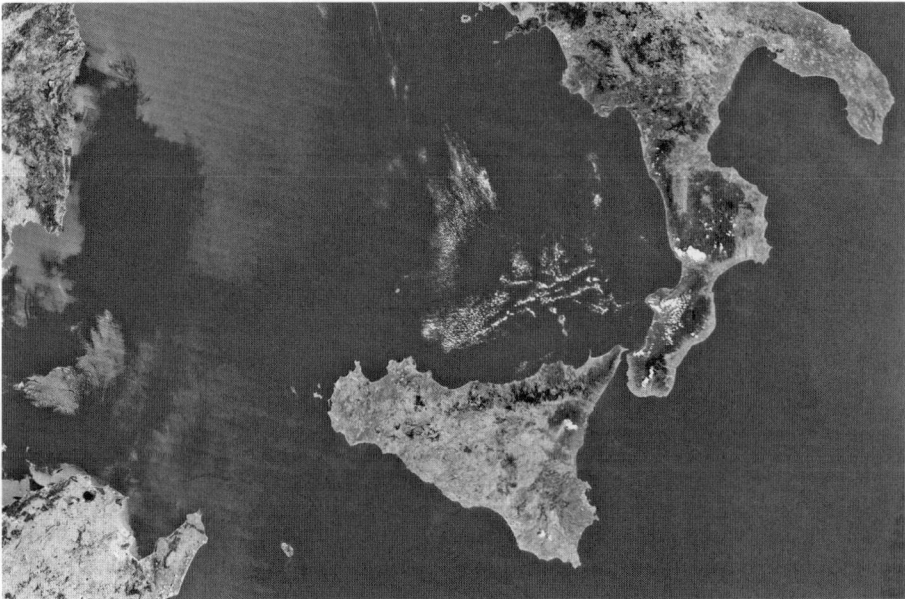

Figure 4.6   MODIS image. On 21 July 2006, the Moderate Resolution Imaging Spectroradiometer (MODIS) flying onboard NASA's Terra satellite captured this image as Sicily's Mount Etna volcano emitted a faint plume of volcanic ash that blew away from the summit, towards the southwest. NASA image courtesy the MODIS Rapid Response Team at NASA GSFC. The MODIS Rapid Response Team provides daily images of Mount Etna

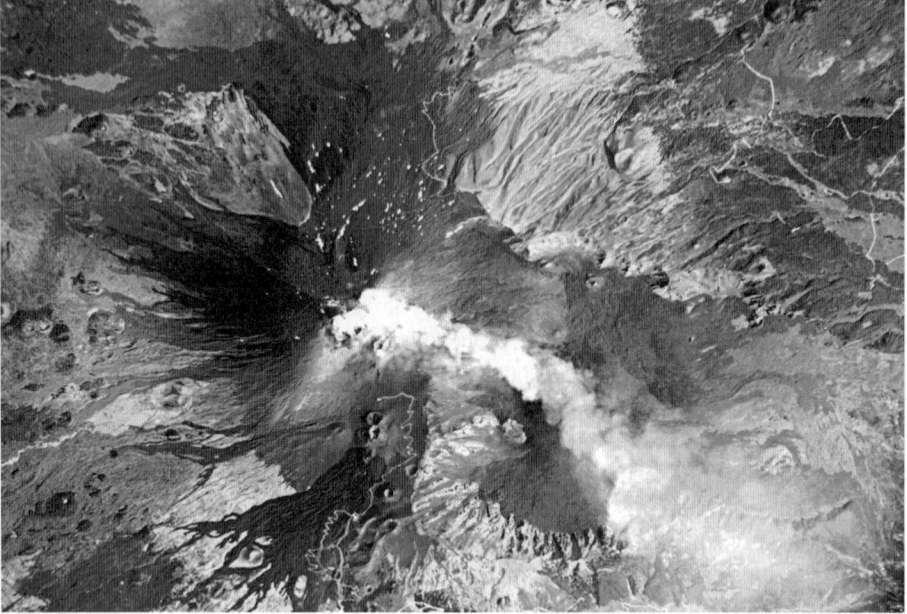

Figure 4.7   Mount Etna. Astronaut photograph ISS013-E-62714 was acquired on 2 August 2006 with a Kodak 760C digital camera using an 800 mm lens, and is provided by the ISS Crew Earth Observations experiment and the Image Science & Analysis Group, Johnson Space Center (http://earthobservatory.nasa.gov/). (*A full colour version of this figure appears in the colour plate section*)

northeastern parts of its volcanic apparatus (Acocella *et al.*, 2003]. Following the eruptions of 2001 and 2002–03, several new eruptive centres along the southern and eastern slopes of the north were created, from which escaped significant volumes of pyroclastic material (Behncke and Neri, 2003; Andronico *et al.*, 2005). The ash and lapillis formed a covering stretching entirely to the top of the volcano, giving it the appearance of a large and homogeneous dark surface. The recent eruptions have also resulted in substantial changes to many of the flows on Mount Etna, especially those closest to the new eruptive centres.

The climate of the Mount Etna area is characterized by an annual mean temperature of 13–18°C and annual mean rainfall of 800–1400 mm. The climate is therefore classified as Mediterranean and/or Mountain Mediterranean. These conditions cause soils to have a moisture regime ranging from Xeric to Udic and a temperature regime ranging from Thermic to Mesic.

The soils of Mount Etna vary largely taxonomically, depending on the mineralogical composition of the volcanic bedrock, land cover and on morphological aspects such as the inclination of slopes. Soils on flat surfaces of the basalt lava plateaus often show a deep and poorly differentiated profile. They are mainly classified as

Eutric Regosols and Eutric Cambisols or, where clay and iron oxide accumulations are present, as Haplic and Chromic Luvisols. Soils generally deriving from volcanic rocks range from Eutric and Dystric Cambisols to Humic Umbrisols. Soils belonging to tephra and acid effusive rocks are mainly classified as Haplic, Humic, Vitric and Silic Andosols.

4.9 Conclusion

There is no such thing as a unique Mediterranean mountain landscape, but rather a complex system of forms and geomorphological processes that reflect the close link with the evolution of Mediterranean structural geology. The close spatial variability of morphological units breaks down the mountain area into very different environments that contribute to the construction of a landscape full of interest and value. The soil component of mountain landscapes reflects the extreme conditions to which these environments have been subject both in terms of climate and geomorphology. The presence of adequate land cover has encouraged the development of advanced and moderately deep soils from different substrates constituting the pedogenic origin. Changes in land cover related mainly to human presence have a strong influence on the conservation of this resource, which is increasingly exposed to erosion and degradation.

References

References marked as bold are key references.

Acocella, V., Behncke, B., Neri, M. and D' Amico, S. (2003) Link between major flank slip and eruption at Mt Etna (Italy). *Geophysical Research Letters* 33:L17310.

Andronico, D., Branca, S., Calvari, S. *et al.* (2005) A multi-disciplinary study of the 2002–03 Etna eruption: insights into a complex plumbing system. *Bulletin of Volcanology* 67:314–330.

Behncke, B. and Neri, M. (2003) Cycles and trends in the recent eruptive behaviour of Mount Etna (Italy). *Canadian Journal of Earth Sciences* 40:1405–1411.

Beydoun, Z.R. (1988) *The Middle-East: Regional Geology and Petroleum Resources*. London: Scientific Press, 292 pp.

Birot, P. (1965) Critères des déformations tectoniques quaternaires (spécialement dans le monde méditerrané). *Revue de Géologie Dynamique et de Géographie Physique* 7:185–195.

Branca, S. and Del Carlo, P. (2004) Eruptions of Mt Etna during the past 3,200 years: a revised compilation integrating the historical and stratigraphic records. In: Bonaccorso, A., Calvari, S., Coltelli, M., Del Negro, C. and Falsaperla, S. (eds), *Mount Etna Volcano Laboratory*. AGU (Geophysical monograph series), vol. 143, pp. 1–27.

Demek, J., Gams, I. and Vaptsarov, I. (1984) Balkan Peninsula. In: Embleton, C. (ed.), *Geomorphology of Europe*. New York: John Wiley & Sons, Ltd, pp. 374–386.

Diaz de Federico, A. and Puga, E. (1976) Estudio geológico del complejo de Sierra Nevada entrelos meridianos de Lanjarón y Pitres. *Tecniterrae* 9:1–10.

Dubertret, L. (1955) *Carte géologique du Liban au 1/200000 avec notice explicative*. Beirut: Republique Libanaise, Ministère des Travaux Publiques, 74 pp.

Dubertret, L. (1975) Introduction à la carte géologique au 1/50000 du Liban. *Notes et Memoires sur le Moyen-Orient* 23:345–403.

Edgell, H.S. (1997) Karst and hydrogeology of Lebanon. *Carbonates and Evaporites* 12:220–235.

Filipovski, Đ. (1996) *Soil of the Republic of Macedonia, Vol II*. Skopje: Macedonian Academy of Sciences and Arts, pp. 175–209.

Gomez, A., Sanchez, S., Simón, M., Salvador, F. and Esteban, A. (1992) Sıntesis de la morfologıa glaciar y periglaciar de Sierra Nevada (Espana). In: Lopez Bermudez, F., Conesa Garcia, C. and Romero Dıaz, M.A. (eds), *Estudios de Geomorfologıa en España*. Murcia: Soc. Española de Geomorfologıa, pp. 379–392.

Jenny, H. (1941) *Factors of Soil Formation*. New York: McGraw-Hill, 281 pp.

Jenny, H. (1980) *The Soil Resource: Origin and Behaviour Ecological Studies*, Vol. 37. New York: Springer.

McFadden, L.D. and Knuepfer, P.L.K. (1990) Soil geomorphology: the linkage of pedology and surficial processes. *Geomorphology* 3:197–205.

Nader, F.H. (2000) Petrographic and geochemical characterization of the Jurassic-Cretaceous carbonate sequence of the Nahr Ibrahim region, Lebanon. MSc dissertation, American University of Beirut, Lebanon, 227 pp.

Nader, F.H., Swennen, R. and Ottenburgs, R. (2003) Karst-meteoric dedolomitisation in Jurassic carbonates, Lebanon. *Geologica Belgica* 6:3–23.

Platt, J.P., Anczkiewicz, R., Soto, J.I., Kelley, S.P. and Thirlwall, M. (2006) Early Miocene continental subduction and rapid exhumation in the western Mediterranean. *Geology*, 34: 981–984.

Quigley, M.C., Sandiford, M. and Cupper, M.L. (2007) Distinguishing tectonic from climatic controls on range-front sedimentation. *Basin Research* 19:491–505.

Raimondi, S., Baldaccini, P. and Madrau, S. (1995) Il clima e il pedoclima dei suoli della Sardegna durante gli anni 1951-80. Atti del Convegno annuale S.I.S.S. Il Ruolo della Pedologia nella Pianificazione e Gestione del Territorio, Cagliari 6–7 giugno 1995, pp. 299–308.

Renouard, G. (1955) Oil Prospects of Lebanon. *American Association of Petroleum Geologists Bulletin* 39:2125–2169.

Riley, C. and Moore, J.McM. (1993) Digital elevation modelling in a study of the neotectonic geomorphology of the Sierra Nevada, southern Spain. *Zeitschrift für Geomorphology* 94:25–39.

Saint-Marc, P. (1974) Etude stratigraphique et micropaléontologique de l'Albien, du Cénomanian et du Turonien du Liban. In: *Notes et mémoires sur le Moyen-orient – Tome XIII*. Paris/Beirut: CNRS, p. 402.

Saint-Marc, P. (1980) Le passage Jurassique-Crétacé et le Crétacé inferieur de la region de Ghazir (Liban central). *Geologie Mediterrenéenne* 7:237–245.

Sánchez, S. (1990) Aplicacıon del estudio de suelos a la dinamica del Rıo Lanjarón. Relaciones suelo-geo-morfologıa. Doctoral thesis, University of Granada.

Sanchez, S., Simon, M., Garcıa, I. and Gomez, A. (1988) Morfogenesis de un sistema nival de Sierra Nevada: Laguna Seca (provincia de Granada). *Cuaternario y Geomorfología* 2:99–105.

Silva, P.G. (1994) Evolución geodinámica de la depresión del Guadalentín desde el Mioceno superior hasta la Actualidad: Neotectónica y geomorfología. PhD Dissertation, Complutense University, Madrid.

USDA (1999) *Soil Taxonomy: A Basic System of Soil Classification for Making and Interpreting Soil Surveys*. United States Department of Agriculture. Agriculture Handbook Natural Resources Conservation Service Number 436.

Walley, C.D. (1997) The lithostratigraphy of Lebanon: a review. *Lebanese Scientific Research Reports* 10:81–108.

Walley, C.D. (2001) The Lebanon Passive Margin and the Evolution of the Levantine Neo-Tethys. In: Ziegler, P.A., Cavazza, W., Robertson, A.H. and Crasquin-Soleau, S.D. (eds), *Peri-Tethyan Rift – Wrench Basins and Passive Margins IGCP 369 results.* Peri-Tethys Mémoire 6. Paris: Mémoires du Muséum National d'Histoire Naturelle.

Woodward, J.C. (2009) *The Physical Geography of the Mediterranean.* **New York: Oxford University Press.**

Yaalon, D.H. (1997) Soils in the Mediterranean region: what makes them different? *Catena* **28:157–169.**

5

Climate and hydrology

Carmen de Jong, İbrahim Gürer, Alon Rimmer, Amin Shaban
and Mark Williams

5.1   Introduction

State-of-the-art reviews on the dynamics of climate and hydrology in Mediterranean
mountain regions show that this is a subject that receives little attention compared to
mountain studies in other, more popular mountain ranges such as the Alps, Rockies
and Himalayas (Messerli and Ives, 1997; Böhm *et al.*, 2001; Viviroli *et al.*, 2003;
Corripio and de Jong, 2005; de Jong *et al.*, 2005a,b,c; Bales *et al.*, 2006). However,
in the context of the significance and contributing potential of the world's mountains
as water towers to the forelands, Mediterranean mountains are classified as having
a high contributing potential to dry lowlands (Viviroli *et al.*, 2007). Climatological
and hydrological literature on the mountainous catchments of the Mediterranean
Basin itself is much sparser than that of other regions of the world with a Mediterranean climate, such as California.

The Mediterranean Basin has unique geographical and climatological characteristics, since its periphery is almost entirely surrounded by mountains at a short
distance from the sea (Figure 5.1). Therefore the dominant climate and hydrology
around the Mediterranean, although essentially determined by latitude, is strongly
influenced by the mountainous relief (Rhanem, 2008). Topographical gradients
from sea level up to 2000–4000 m altitude are not uncommon. Mediterranean mountains zones, whether consisting of chains or individual peaks, act locally as cold
islands with strong climatic gradients (Ozenda, 1975). Typical for such extremes
is, for example, the High Atlas in Morocco, which can experience temperatures of
−18°C yet is surrounded by the Sahara within less than 200 km distance.

Mediterranean climatology is complex, and basically unites three different zones:
the sub-Saharan climate, the dry continental climate to the east and the more continental climate to the north. It forms the boundary between the mid-altitude and tropical climate, between vegetated and non-vegetated, and between highly and poorly
developed regions. The climate, together with the distribution and altitudinal range
of mountain ranges (Table 5.1), strongly influences water availability and ensuing

Mediterranean Mountain Environments, First Edition. Edited by Ioannis N. Vogiatzakis.
© 2012 John Wiley & Sons, Ltd. Published 2012 by John Wiley & Sons, Ltd.

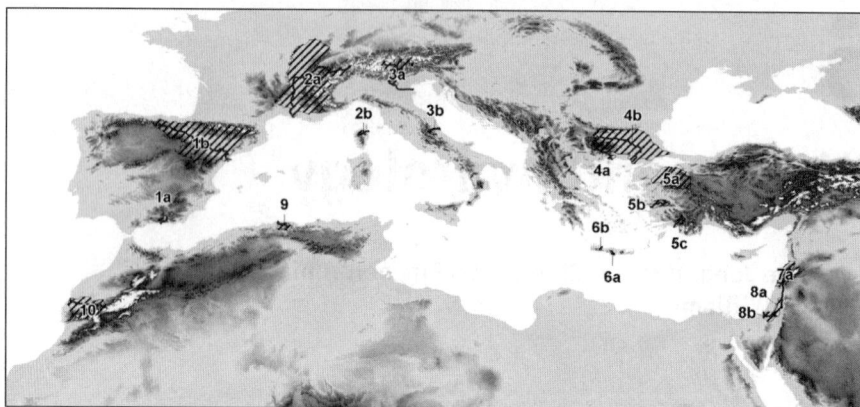

Figure 5.1 Mediterranean mountain river basins (including mountain ranges on mainland and islands) of pilot sites corresponding to Table 5.2. (*A full colour version of this figure appears in the colour plate section*)

river regimes. Depending on altitude, some of these are dominated by snowmelt, others are purely rainfed. Snowmelt discharge regimes are generally more reliable, in particular where snow acts as a buffer and releases run-off with a delay in spring-time. Nevertheless, snowmelt volumes are influenced by temperature (Cazori and Dalla Fontana, 1996) and can be restricted by sublimation. Relationships between rainfall run-off regimes are not always simple either, especially at the catchment level, where evapotranspiration can have important influences (Latron *et al.*, 2008). The seasonal temperature evolution in Mediterranean regions has a direct effect on evapotranspiration dynamics, which, in combination with the seasonal dynamics of rainfall, cause the succession of wet and dry or very dry periods during the year (Latron *et al.*, 2009). This seasonality of the Mediterranean mountain climate has a

Table 5.1 Classification of Mediterranean vegetation levels and equivalent thermal thresholds. After Akman and Daget (1971), Ozenda (1975), Quezel (1976), Peyre (1979), and Rhanem (2008)

Vegetation and thermal thresholds	Thermal limits (°C)	Classification
Oro-Mediterranean	<-15	Glacial
	-15 to -12	Extremely cold
	-12 to -9	
	-9 to -6	
Mountain Mediterranean	-6 to -3	Very cold
Superior and supra-Mediterranean	-3 to 0	Cold
Meso-Mediterranean	$0 < 3$	Cool
Thermo-Mediterranean	3 to 7	Temperate
	7 to 10	Hot
	>10	Very hot

strong influence on the spatio-temporal dynamics of both soil moisture (Menziani *et al.*, 2005) and the groundwater table. Whilst literature on precipitation/run-off and evapotranspiration in the Mediterranean lowlands is abundant, studies on snow and glacier hydrology in Mediterranean mountains are generally lacking and work is mostly confined to the Pyrenees (López-Moreno and Garcia-Ruiz, 2004).

Mountains in general comprise one of the most sensitive environments of our globe and are therefore excellent indicators of but also victims of climate change. This concerns not only the Mediterranean mainland and island mountains but equally the Alps. Mountains act as amplifiers for different climatological phenomena related to altitude, such as radiation, temperature, precipitation and relative humidity. Special valley effects in winter, such as inversions, account for additional perturbations in the system. The spatial variability of these parameters is also amplified in mountains, bringing more extreme conditions such as flooding (Bacchi and Villi, 2005) or water scarcity (de Jong, 2009).

As for other mountain chains worldwide, Mediterranean mountains are generally considered as the 'water towers' for the surrounding lowlands (Viviroli *et al.*, 2003). In the semi-arid areas of the Mediterranean, run-off can contribute from 5% to 90% of total water supply (Viviroli *et al.*, 2007). Similarly, these mountains are classed among the world's most important biodiversity hotspots and are therefore particularly sensitive to climate and human change. The Mediterranean is of special importance since it is a region with its own endemic species (citrus fruit, olive trees, almond trees, *Pinus pinea* etc) whose agricultural exploitation has long been influenced by the prevailing climate.

The main challenges of the Mediterranean mountain climate and hydrology are its extremes, both in terms of droughts and floods, man-made modifications to the water cycle, in particular dams, and their amplification by climate change. The climatic regime together with the influence of dams and other man-made structures on the main rivers, cause strong variations in freshwater input into the Mediterranean Basin (Ludwig *et al.*, 2009). Human water consumption is greatest along the Mediterranean coasts, where population pressure is strongest, due to permanent and tourist populations and to intensive irrigation. The upstream-downstream relations for water availability and consumption are particularly important.

The analysis of past climates, for example from lake sediments, is very important for future climate projections (Tonkov *et al.*, 2002; Ghosn *et al.*, 2010). The Mediterranean region has always been a very sensitive environment, and indicators for climate change trace back as far as the Bible and the Pharaonic Period (droughts, floods, pests, famines). During the Holocene, there have been clear triggers and responses to changes in climate and hydrology, reflected by the spread in farming, and maritime and terrestrial trade around the Mediterranean and Africa. Historically, coastal zones have dominated and the importance of their mountains in the hinterland has often been neglected.

The Mediterranean is particularly sensitive to climate change (Jeftic *et al.*, 1992). Climate change and increasing population pressure have caused water stress in many mountain regions that depend on rainfall and meltwater from snow and glaciers. Few glaciers remain, mainly in the Mediterranean Alps and the Apennines. Fluctuations in precipitation, shifts in meltseason, water abstraction for artificial snow and dam reservoirs all affect the water cycle, especially river flow. This in turn has strong impacts on drinking water, tourism, irrigation, agriculture and industrial production in downstream regions. Since existing environmental and socioeconomic problems in mountain environments are complex and confined, these can be significantly accelerated and amplified by climate change (de Jong and Schoeneich, 2009). For all latitudes, climate change scenarios predict severe temperature increases with elevation in mountains (Bradley *et al.*, 2004). Within the natural environment, the 0°C isotherm is critical for the persistence of snow and glacier resources, therefore mountains are particularly susceptible to climate change (Diaz, 2004). Mountains have less buffering capacity to withstand the impacts of climate change. Since the biota is already at the edge of tolerance in mountains, it requires little impact to cause a major disequilibrium. The same is true for the social and economic environment, where ski areas and snowline altitudes are highly sensitive to the length and intensity of snowfall, especially on south-facing slopes of the mountains. Increases in temperatures and decreased snowfall (but possible increases in rainfall) as predicted in climate change scenarios can easily erode the existence of traditional, commercialized winter tourism, as described in the recent OECD report (Elsasser and Messerli, 2001; OECD, 2007). Means of economic subsistence, including agricultural activities such as dairying or forest harvesting, are also highly sensitive to changes in the physical environment. Such activities can, however, be menaced by the increasing natural hazards and risks associated with the ongoing decay of glaciers and permafrost.

In addition, the alteration of water flows in terms of water quantity has important impacts on water quality through changes in concentrate mixing levels, water velocity, temperatures and oxygenation levels. Reduced water flows generally increase pollution levels by increasing pollutant concentrations, in particular during the drier and warmer seasons. This has impacts on the health of both ecosystems and humans, while simultaneously posing an economic burden for restoration strategies.

5.2   Climate and physical characteristics

The Mediterranean Basin is representative of a typical Mediterranean climate with dry summers and wet winters (Rivas-Martínez *et al.*, 2001). Around the Mediterranean there are 16 important mountain regions influenced by the cryosphere, in particular snow, creating important snow-fed regimes. These include both continental and island ranges, some of which are summarized in Table 5.2. There is a large variation in: (i) altitude (from 1000 to 4800 m); (ii) climate systems affecting snow and glacier accumulation; (iii) snow cover duration (from 2 to 10 months per year); (iv) geology and geomorphology and therefore different surface and groundwater discharge regimes; (v) distance from source to sea; and (vi) values of downstream agriculture and tourism. The Platanias and Almyros rivers on the Greek island of

Table 5.2 Physical and socioeconomic characteristics of Mediterranean mountain river basins under the influence of snow and glacier melt including different mountain chains, countries, river basin areas, distance to sea, maximum altitude, maximum altitude of basin and meteorological stations, annual rainfall and snowfall, discharge and average population density. The 10 MountSnowMed (2007) pilot sites are marked as shaded pilot sites. Please note that most of the data for snowfall are based on estimations only

Pilot site	Mountain chain	Source country	River	Approx. basin area (km²)	Approx. distance source to sea (km)	Max. altitude of basin (m)	Highest altitude meteorological station (m)	Approx. annual rainfall (mm)	Approx. annual snowfall (mm)	Specific discharge (L/s/km²)	Agricultural water use: % of local water use	Population density (/km²)
1a	Sierra Nevada	Spain	Guadalfeo	1300	100	3481	2502	1370	1200	nd	nd	69
1b	Pyrenees	Spain	Ebro	85 500	580	3404	2202	622	124	19.00	>35	33
2a	Alps	France	Rhône	94 300	690	4807	3000	900	2500	17.60	49	118
2b	Corsica	France	Golo	1000	90	2706	2360	600	1300	15.30	49	100
3a	Alps	Italy	Adige	12 000	410	3899	3325	765	135	19.60	nd	136
3b	Apennines	Italy	Vomano	800	70	2912	1000	1102	500	17.41	60	50
4a	Balkan	Bulgaria	Mesta-Nestos	5800	230	2925	2392	830	nd	nd	93	31
4b	Balkan	Bulgaria	Ergene Maritza	14 600	180	1018	300	604	2500	2.90	72	75
5a	Mount Uludag	Turkey	Susurluk	30 000	320	2543	1920	712	2000	7.20	72	95
5b	Mount Kazdaglari	Turkey	Karamenderes	3200	120	1174	nd	650	nd	5.30	77	63

(Continued)

Table 5.2 (*Continued*)

Pilot site	Mountain chain	Source country	River	Approx. basin area (km²)	Approx. distance source to sea (km)	Max. altitude of basin (m)	Highest altitude meteorological station (m)	Approx. annual rainfall (mm)	Approx. annual snowfall (mm)	Specific discharge (L/s/km²)	Agricultural water use: % of local water use	Population density (/km²)
5c	Taurus	Turkey	Dalaman	5100	230	2420	nd	876	3000	12.40	70	189
6a	Lefka Ori	Greece	Platanias River	500	20	2453	1050	1980	2000	nd	85	nd
6b	Pseilortis Mountains	Greece	Almyros River	200	20	2456	1450	1640	2000	nd	82	nd
7a	Lebanon	Lebanon	15 Lebanese rivers	7500	130	2850	1900	1200	nd	nd	67	234
7b	Anti-Lebanon	Lebanon/ Israel	Jordan	800	70	2814	1450	960	650	nd	80	250
8a	Mount Ebal	Palestine	Mount Ebal River	3200	30	1027	1000	600	50	nd	53	330
8b	Mount Hebron	Palestine	Palestinian river	4200	150							
9	Tell Atlas	Algeria	Oued Sebaou	2400	70	2305	1950	1000	nd	nd	nd	354
10	High Atlas	Morocco	Tensift	20500	240	4167	3200	600	1000	0.34	75	98

Crete are marked by some of the highest energy gradients, draining in only 20 km from the highest peak (at nearly 2500 m) to the sea. Other mountainous islands such as Corsica and the near coastal range of the Apennines follow similar trends, draining from 2900 m altitude to the sea in only 70 km.

Although water resources and related issues are relatively well known and investigated in Mediterranean coastal zones, monitoring and modelling of snow-precipitation inputs in the mountains and hinterlands as well as their cascading effects still remain more or less a grey area. In the Mediterranean mountain regions, the shortage of studies is even more serious due to the dispersed nature or total lack of monitoring networks, the limited focus and attention given to snow and glacier-related research, the remoteness and inaccessibility of some sites, the difficulties in maintenance of a meteorological station at higher altitudes (Shaban *et al.*, 2005; Rimmer and Salingar, 2006) and the limited technological development of meteorological and hydrological services in some countries. Often monitoring stations with long-term data are lacking, so that it is difficult to calculate long-term trends or impacts of climate change (Figure 5.2). Some mountain ranges are devoid of long-term measuring stations (e.g. the High Atlas, Pyrenees, Sierra Nevada). In general there is a greater density of monitoring stations in the northern Mediterranean, and long-term data (since the 1850s) are derived almost exclusively from the western and northern Mediterranean; monitoring has been established in the southern Mediterranean predominantly since the beginning of the twentieth century, and in the eastern Mediterranean only from the 1940s (Xoplakie *et al.*, 2004).

Remote-sensing techniques to estimate volumes of water originating from snowmelt are common (Shaban *et al.*, 2004; Chaponnière *et al.*, 2005; Boudhari

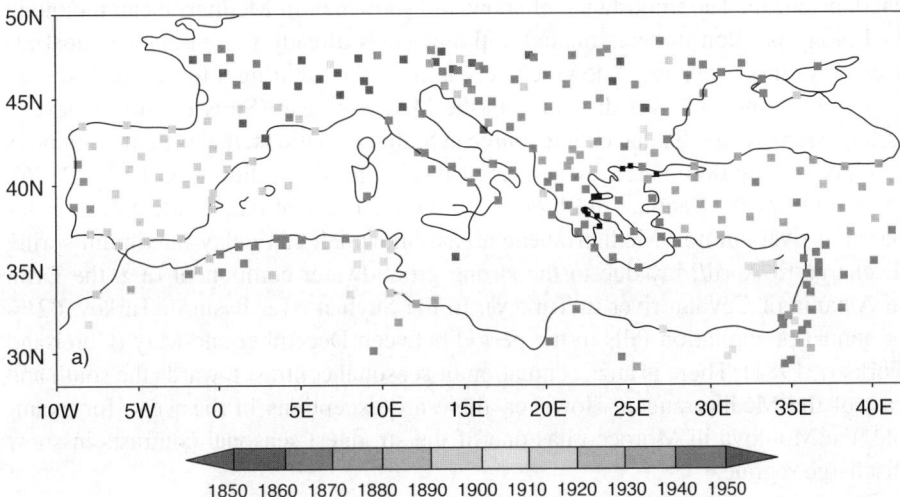

Figure 5.2 Dates of beginning of precipitation records (in decades) across Mediterranean. Reproduced from Xoplaki *et al.* (2004), with permission. (*A full colour version of this figure appears in the colour plate section*)

et al., 2009) but limited to certain Mediterranean mountain zones. It is important in future to improve the density and standards of snow-related measuring techniques, including automatic meteorological stations, combined with high-resolution remote sensing, altimetry satellite sensors and digital photogrammetry for monitoring snowpack and estimating snowmelt in addition to 3D modelling techniques.

5.2.1 Rivers

Mediterranean rivers usually have extreme regimes, with very low summer discharge and higher winter discharge reflecting the prolonged dry summers and wet winters. The unit discharge of Mediterranean rivers varies considerably between 0.3 and 20 L/s/km^2, according to the geographical and topographical location within the Mediterranean Basin. Discharge can be very erratic, reflecting both floods and droughts, whereby flood flows can be several hundred orders of magnitude higher than normal flow (de Jong *et al.,* 2008). Most river flow is highly regulated through dams and reservoir storage. The existence of ephemeral or seasonal streams and rivers is another special characteristic of Mediterranean hydrology that determines the uniqueness of the relevant ecosystems. Irregular river flow poses particular water resource challenges with respect to the application of the European Union Directives (Skoulikidis, 2009).

5.2.2 Snowmelt hydrology

This chapter will focus mainly on the more neglected snowmelt regimes of the Mediterranean. The importance of snow and snowmelt in Mediterranean mountain hydrology is often underestimated, although it is already the parameter most affected by climate change. Snowmelt acts as an important trigger for floods and is an major contributor to river discharge in the Mediterranean (Strasser and Etchevers, 2005). Many rivers that have their source at higher elevations, that is in mountainous catchments, are dominated by snowmelt discharge. According to CIESM (2006), the ratio of peak to annual discharge is often one order of magnitude greater compared to rivers in non-Mediterranean areas. Snowmelt can delay maximum spring discharge to April/May due to the strong groundwater component (e.g. the Drini in Albania or Ceyhan river in Turkey). In the Seyhan river basın in Turkey, 62 % of annual precipitation falls in the period between December and May (Gürer and Türksoy, 1981). There is an accentuation of seasonal contrast towards the south and east of the Mediterranean. However, there are exceptions in the west; for example, the Moulaya in Morocco has one of the strongest seasonal contrasts in snow discharge regime.

The snowmelt contribution to discharge (estimated at 25–30%) plays an important role in many major Mediterranean rivers yet precise quantification of snow and glacier-melt contribution to discharge is still relatively poor (see Table 5.2). This preliminary table (Table 5.2) was compiled by the authors and it shows that

Figure 5.3 Irrigated fields dependent on snowmelt regime in the Oued Yagour River Basin in High Atlas, Morocco (photo: Simmoneaux). Reproduced from Ludwig *et al.* (2009), with permission

significant improvements are required to establish precise data for snowfall/rainfall precipitation and discharge. The lack of literature on Decision Support Systems for water resources originating from mountains is almost certainly due to the poor quality and quantity of such input data. The same problem applies to the estimation of the relative contribution of snowfall to the annual water budget and groundwater. For example, it is estimated that 12% of the Hermon aquifer recharge below Mount Hermon (Israel) is replenished by snowmelt (Rimmer and Salingar, 2006). In the Tensift catchment of the High Atlas, Morocco, more than 60% of the specific annual discharge of only 0.34 L/s/km^2 is derived from snowmelt (Figure 5.3). More than 75% of the total discharge is used for irrigation of agricultural land in this catchment with an average population density of 98 persons/km^2. In contrast, the Dalaman catchment in Turkey has a specific annual discharge that is 36 times higher, at 12.4 L/s/km^2, of which snowmelt accounts for 40–50%. Again, more than 70% of this low specific discharge is used for irrigation agriculture in a catchment that has nearly double the population density with 189 persons/km^2. On the whole, Turkey, Algeria, Morocco and Italy have the highest percentage of mountain populations in the Mediterranean. For some of the most densely populated mountain basins, such as the Jordan in Lebanon and Oued Sebaou in Algeria, precise snow-related discharge data are lacking. Nearly all mountainous catchments that use more than 70% of their total water for agriculture depend on some of the lowest specific discharges and highest input from snowmelt, thereby rendering them particularly vulnerable even for 'business-as-usual' climate and anthropogenic scenarios. These data require improvement in future projects through various monitoring techniques.

5.2.3 Snow and water quality

Apart from the quantity of water, the quality of water derived from snow is of major importance. Thus the chemical and radiochemical quality of snow requires detailed examination (Villa *et al.*, 2006). Agricultural fertilizers and other chemical substances that were used in past decades and have since been prohibited were transported in the atmosphere, then condensed, precipitated and concentrated in snow at higher altitudes. With rapid temperature increases due to global warming in recent years, snow is melting faster than usual including the older snow layers. Thus the polluted substances that were 'locked' in the snow are re-emerging over short time periods in the meltwater flow with unusually high concentrations. In addition, very high levels of radioactivity can be measured in the snow and in the groundwater derived from snowmelt in the Alps and other mountain regions such as the Apennines. Because the Chernobyl accident of May 1988 was associated with widespread snowfall or rainfall in the Alps, the radioactivity levels at high altitudes are to this day persistently high. However, the spatial distribution of radioactivity is much more heterogeneous in the mountains than on the plains. The release of such substances through ablation of the older snow cover can substantially influence water quality and limit the availability of water for human consumption.

5.2.4 Ecohydrology

Snow- and rainfall-derived water quality and quantity have important effects on aquatic ecology, such as streams and lakes, since temperature and substrate stability strongly influence macroinvertebrate community structure in alpine snow- and glacier-fed streams (Castella *et al.*, 2001). Aquatic ecology is much more vulnerable than terrestrial ecology to physical impacts. According to Ricciardi and Rasmussen (1999), aquatic faunas are five times more vulnerable to extinction than terrestrial faunas. Literature on this topic is sparse and confined to the Alps but it is necessary to extend this field of study and ensure that attention is given to the relations between aquatic biodiversity and economic values (such as water quality and tourism).

Mediterranean mountain forests are particularly dynamic ecosystems that influence the hydrological cycle in terms of evapotranspiration, condensation, discharge and snow retention. Forest and forest watersheds play an essential role in sustaining and protecting water supplies (FAO, 2004;, Garcia-Santos *et al.*, 2005). However, mountain biodiversity has been strongly negatively affected by the ongoing trends in agricultural decline (Mitchley *et al.*, 2006). The change in forest composition over the last thousands of years from deciduous and coniferous trees to forest patches dominated by hard-leaved shrubland has modified the regional hydrology and climatology. Intensive utilization of forest by humans often overrides the climatic effects. However, climate change may have a considerable impact on changes in plant species composition – in the literature these are usually related to temperature and CO_2 concentrations. Climate change may enhance the spread of wildfires and result in species recomposition. It is necessary to advance the issue of interactions between forest and changing hydrology under scenarios of climate change as well as interactions between ecosystems, droughts and wildfires, taking into account human interactions.

5.2.5 Dams and hydrology

Dam reservoirs have a major influence on river flow, particularly where water is abstracted for irrigation (Figure 5.4). The Ebro in Spain, for example, has experienced an estimated 63% discharge reduction, approximately half the reduction being attributed to climate change and the other half to the impacts of the dam. Similarly, the Moulouya River in Morocco has experienced a more than 70% decrease in discharge since multiple dam construction.

It is estimated that discharge into the Mediterranean Sea has been reduced by at least 50% over the past 100 years due to a combination of climate change and anthropogenic use (Ludwig *et al.*, 2003). The influence of dams has been particularly strong in modifying the discharge regimes in mountain environments. In order to optimize the management of water resources with a snow component, it is

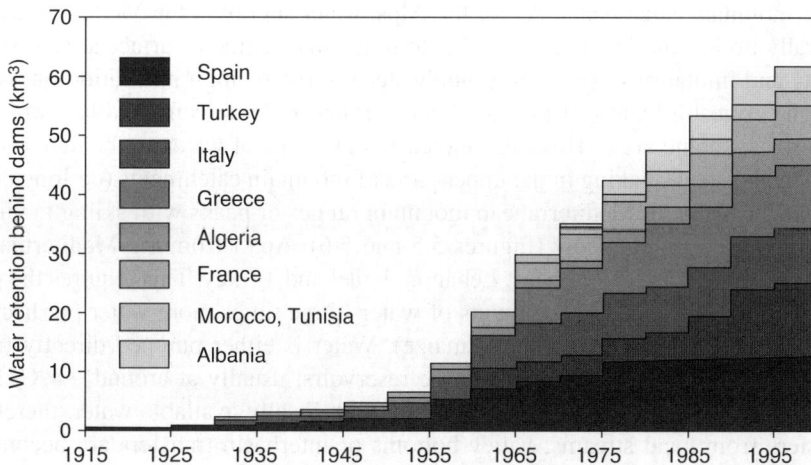

Figure 5.4 Evolution of the water retention behind dams in the drainage basin of the Mediterranean Sea. Reproduced from Margat et Treyer (2004) and Ludwig *et al.* (2009), with permission

important to take into account the possible effects of changing snow seasonality, changing absolute quantities of snowmelt as well as loss of discharge due to increasing river impoundment by dams. Apart from changing discharge regimes, the effects of sedimentation of dam reservoirs will strongly influence the distribution of water resources available for hydropower energy, irrigation, vegetation and human consumption. Demand for electricity through hydropower will increase with a growing population and industry, so that uncertainty about reservoir filling levels will increase in the future. Water demand for agriculture, in particular for export crops, is also expected to grow. Possible earlier snowmelt occurrence will reduce the availability of water during the summer period when water demands are particularly high. The consequent increased sediment entrapment in the dams will have significant impacts on the sediment and nutrient transport to the deltas, enhancing coastal erosion and loss of wetlands. Parallel to this, export of virtual water (the volume of freshwater used to produce a product, measured at the place where the product was actually produced – Hoekstra, 2003) from the Mediterranean region is likely to increase, for citrus fruits and dates for instance, whilst the regional differences will possibly become larger (Hoekstra, 2003). Moreover, peaks in energy and water consumption by tourism during the spring and summer will add to the burden on the available water resources. Since the highest water consumption is for agriculture, it is important to establish the dependencies of snowmelt discharge, rainfall-runoff and agricultural practices for different basins in the future.

5.2.6 Tourism and hydrology

Water supply and tourism is an issue of concern not only in the coastal fringes of the Mediterranean (for hotels, swimming pools, golf courses, etc.) but also increasingly in the mountain catchments. As for the Alps, water supply in the Mediterranean is naturally limited at higher altitudes due to limits in catchment surface area, discontinuity and limitations in size of groundwater reservoirs, high infiltration rates due to high permeability, low discharge of zero- or first-order streams and lack of natural surface storage areas. However, increasing amounts of water are being used for tourism and snow-making in the upper parts of mountain catchments (de Jong *et al.* 2009a). Virtually all Mediterranean mountain ranges or peaks with skiing facilities now produce artificial snow (Figures 5.5 and 5.6). Apart from the Mediterranean Alps, these include the Pyrenees, Lebanon, Israel and Turkey. This requires the permanent availability of large amounts of water (three times more water per hectare than that used for the irrigation of maize). Water is either pumped directly from streams or stored in large water storage reservoirs, usually at around $+4°C$. It is increasingly difficult to fill these reservoirs with locally available water, therefore transfers from local streams, valley bottoms or interbasin transfers are becoming more common. Such transfers do not take place without impact on the local ecology, hydrology and soils. In the source areas where water is captured it causes a water deficit, whereas in the destination areas a water surplus arises. Some resorts in the

Altitude (m)

☐ > 2 000
■ > 1 600
▨ > 1 200
▨ > 800
☐ > 400
☐ > 0

🎿 artificial
snow production

Figure 5.5 Regions with artificial snow production in the Mediterranean mountain regions. Size of symbol corresponds roughly to amount of snow production closely related to relative water volume abstracted. Reproduced from Carmen de Jong (2007), with permission. (*A full colour version of this figure appears in the colour plate section*)

Figure 5.6 Water storage reservoir for artificial snow making on Mount Hermon. Note problems of sedimentation indicated by colour of water and deposits along shoreline. http://en.wikipedia.org/wiki/Mount_Hermon. Reproduced from Xoplaki *et al.* (2004), with permission.

Pyrenees are even considering collecting rainwater or snowmelt from the roofs of public buildings in order to supply one-third of their mountain reservoirs. In terms of climate change, the OECD (2007) has reported that about 50% of Mediterranean ski areas will close with a temperature increase of 2–4°C.

These rapid and diverse developments in water use require profound reflections of water sharing and water management systems to avoid water conflicts and even more water shortage. Therefore it is important to consider the development of decision support systems to economize, conserve and replenish water resources.

5.2.7   Decision support systems (DSS)

Decision support tools for integrated water resources management in mountain regions are generally not well developed (Leavesley *et al.*, 1996; Schreier, 2003; SAEFL, 2003; FAO, 2004; de Jong *et al.*, 2004, 2005c, 2006, 2008). This is mainly due to the problem of hydrometeorological modelling in mountain catchments and in particular the lack of validation data for models (Leavesley *et al.*, 1996). In Mediterranean mountain regions the development of DSS with close stakeholder participation is an essential issue in the immediate future (SAEFL, 2003; FAO, 2004; de Jong *et al.*, 2006). Thus, in future it will be necessary to manage available snow- and rainfall-derived water resources in the light of adaptation strategies rather than rely on the more traditional base of water availability (European Environment Agency, 2009). The traditional water economy will no longer be adapted to the growing water demands under increasing environmental and climatic pressures. Conflicts between water for drinking, agriculture, hydroelectricity, artificial snow production and tourism are likely to intensify. It is therefore important to analyze the costs and benefits of adaptation strategies under different scenarios for mountain regions, as was initiated for the Rockies by the Aspen Initiative of the Aspen Global Change Institute (2006).

For this purpose it is essential to develop physical, biological, socioeconomic and political indicators associated with hydrological issues (Table 5.3). Different key criteria associated with a range of different sectors can be identified, which in turn can have a series of impacts in each specific indicator field. For example, within the field of physical indicators, air temperature and seasonality of rainfall/snowfall affect tourism and socioeconomic activities. The key impacts are on the ski season, artificial snow production, number of tourists, water availability (especially drinking water), buffering of summer discharge, aquatic ecosystems, frequency of wildfires and the local economy. Within the field of socioeconomic indicators, key criteria can include, for example, urbanization, diversity and spatial extent of mountain valleys; the socioeconomic sectors affected include industry and tourism, through impacts on roads, railways, housing, flooding, natural hazards and water availability. Governance is another important key criterion, associated with integrated water resources management, negotiation systems for water conflicts, and sustainable and safe territorial management. The sectors influenced include water resources,

Table 5.3 Key physical, biological and socioeconomic criteria identified taking into account different sectors and important impacts

	Key criteria	Sectors	Impacts
Physical	Air temperature and seasonality of rain/snow	Tourism, socioeconomic activities	Ski season, artificial snow production, number of tourists, water availability (especially drinking water), buffering of summer discharge, aquatic ecosystems, frequency of wildfires, economy
	Snowline altitude	Infrastructure, socioeconomic, health	Roads, railways, winter tourism, snow quality (air pollution)
	Glacier size and orientation	Tourism, economic	Number of tourists, buffering of summer flow, vegetation extent, natural hazards
	Permafrost altitude, freeze-thaw cycles	Infrastructure, housing, socioeconomic	Roads, railways, tourist hiking and cycling paths, natural hazards
	Slope gradient and slope processes	Agriculture, infrastructure	Agricultural productivity, infrastructure and natural hazards
	River levels and groundwater discharge	Agriculture/energy/tourism/health	Buffering of droughts, aquatic ecosystems, water quality, sediment transport, import/export of virtual water, energy production levels, rafting tourism, economy
	Natural and artificial reservoirs (dams, etc.)	Economic, agriculture, energy	Mountain irrigation agriculture, aquatic ecosystems, downstream dam impacts, acidification, groundwater levels, biodiversity, river bed stability

(*Continued*)

Table 5.3 (*Continued*)

	Key criteria	Sectors	Impacts
Biological	Aquatic ecology	Tourism, health	Number of tourists, water quality
	State of alpine meadows	Socioeconomic, tourism	Population density and distribution, economy
	Biodiversity	Agriculture, forestry	Quality of mountain honey, quality of wood
	Forest diversity and density	Industry, energy, handicrafts	Forest extent, forest harvesting and quality of sales
Socioeconomic	Emigration and immigration	Socioeconomic, tourism	Population density and distribution, economy
	Seasonality of tourism	Tourism, economic	Housing, hotels and restaurants, retailing, water availability, economy
	Urbanization, diversity and spatial extent of mountain valleys	Socioeconomic, industry, tourism	Roads, railways, housing, flooding, water availability, natural hazards
Governance	Integrated water resources management; negotiation systems in water conflicts; sustainable and safe territorial management	Water resources, territorial planning, risk management, information, participative democracy, educative	Water availability, equity, sustainability, negotiation, prevention, peace, capacity building, information, awareness, participation

territorial planning, risk management, information, participative democracy and education. The impacts include water availability, equity, sustainability, negotiation, prevention, peace, capacity building, information, awareness raising and participation. In this context, the recent EU Water Framework Directive is an essential guiding principle for integrated water resources management at a European level based on the ecological quality concept and on adaptive approaches with strong stakeholder participation.

5.3   Vulnerability of water resources to climate change

Snow-fed mountain regions around the Mediterranean are particularly vulnerable to impacts of climate change due to the combination of intensive utilization of their natural resources, their distribution around the rapidly warming Mediterranean Sea at the northern limit of the Sahara (HYMEX, 2007) and their natural tendency for droughts (Figure 5.7). Glacier-influenced regimes are restricted nowadays to the Alps and one remaining glacier in the Apennines. Globally, the Mediterranean concentrates the largest number of mountain regions with a high water-contributing potential to the neighbouring dry lowlands (Viviroli *et al.*, 2007). Mountain water supply is crucial especially in arid and semi-arid regions where vulnerability to seasonal and regional water shortage is high. However, since Viviroli's approach was developed on a global scale, the scale is still very coarse and requires a higher resolution tuned to the basin scale in order to assign values for individual Mediterranean mountain chains. Furthermore, climate change impact studies in mountainous areas only concentrate on aspects such as snowline altitude, percentage catchment covered by perennial snow, snow depth and snow seasonality. The Mediterranean Alps, such as the Monte Rosa, have a significant influence on irrigation, mainly for rice in the Piedmont region, and glacier retreat as well as damming of rivers for hydropower causes concern for downstream water availability (Alp-Water-Scarce, 2008–2010).

Climate change reviews for mountain regions show that the most significant changes are expected to occur in mountains under the influence of continental and Mediterranean climates (Beniston, 2003). The Mediterranean region is particularly vulnerable to climate change impacts, according to a report by the European Environment Agency (2005, 2009). The most sensitive regions in terms of hydrology include the Mediterranean, Alps, and central and eastern Europe (Watson *et al.*, 1997). In terms of vulnerability and potential impacts on mountains, it is expected that run-off will decrease by 15% in southern Europe, droughts will increase, snow and ice will decrease and water demand will increase, especially in summer. Snow accumulation in most Mediterranean mountain regions is rapid and discontinuous, therefore a continuous snow mantle as experienced in the more humid Alps is often not present (Schulz and de Jong, 2005). Since snow ablation is already highly sensitive to physical factors, any temperature changes associated with climate change will cause even more rapid changes in snow cover distribution.

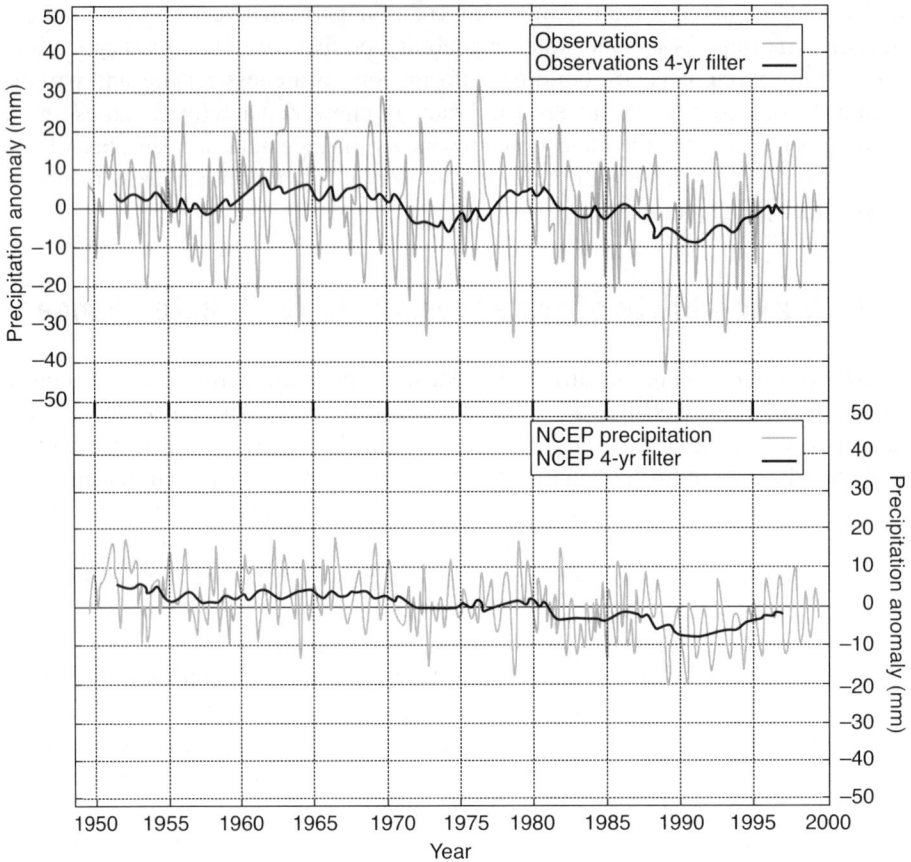

Figure 5.7 Upper panel: wet season precipitation anomalies averaged over the 292 sites (black line) and 4-year low pass filter (red line). Lower panel: wet season precipitation anomalies averaged over NCEP (National Centres for Environmental Prediction) reanalysis data (black line) and 4-year low pass filter (red line). Reproduced from Ludwig *et al.* (2009), with permission

Due to the altitudinal limits of most Mediterranean mountain regions (average altitude 2800 m) they are often already close to the snow and ice limit. Any increase in temperature according to current climate scenarios can significantly threaten the reliability of the snow- and ice-fed resources. The ecological and economic buffering potential of the mountains increases from south to north and from east to west, but the details of the impacts of climate change are still unknown. Since historical times, the Mediterranean mountain water towers have played an unquestioned role in the development of their forelands. Depopulation of the mountains is a major issue that has impacted the present ecology and physical landscape (McNeill, 1992). Many of the downstream communities, predominantly agricultural and touristic, are already under extreme water stress and strongly dependent on highly variable

snow- and glacier-driven river discharge regimes in direct competition with water abstraction by hydroelectric dams. These Mediterranean mountain regions are
therefore particularly vulnerable in terms of providing water resources.

According to Ludwig *et al.* (2009), there was a sharp decrease in precipitation
and average freshwater flux to the Mediterranean Sea over the 40 years from 1960
to 2000. Although data are patchy, precipitation decreased by about 11% both in
the eastern and western Mediterranean Basin. This is equivalent to a total estimated
decrease in precipitation of 62 mm between 1960 and 2000. Temperature increased
markedly, with the biggest increase of 1.3°C being observed in the northwestern
Basin. Higher temperatures are likely to increase evapotranspiration and therefore
decrease discharge. In the Gulf of Lions, water discharge decreased mainly as a
result of temperature-related reduction of basin internal storage in snow, soils and
groundwater reservoirs. The estimated freshwater discharge into the Mediterranean
decreased to a serious extent, by 15–22%, between 1960 and 2000, which is equivalent to a decrease of 80–100 km^3/year (the combined annual discharge of the Rhône
and Po; see Figure 5.8). For the Balkans, there have also been dramatic discharge
reductions, for example a 30% decrease in the Drin between 1965 and 1984, and between 1960 and 2000 a 57% reduction for the Axios, with a catchment of more than
20 000 km^2 (Skoulikidis, 2009). In the eastern as well as the western Mediterranean
Basin, the gap between increasing temperatures and decreasing precipitation has
been widening over the last 40 years (Figure 5.9). Generally, this will result in more
droughts. Considering present trends in global warming and decreases in snowfall,
the security of the late spring/early summer discharge may be seriously threatened

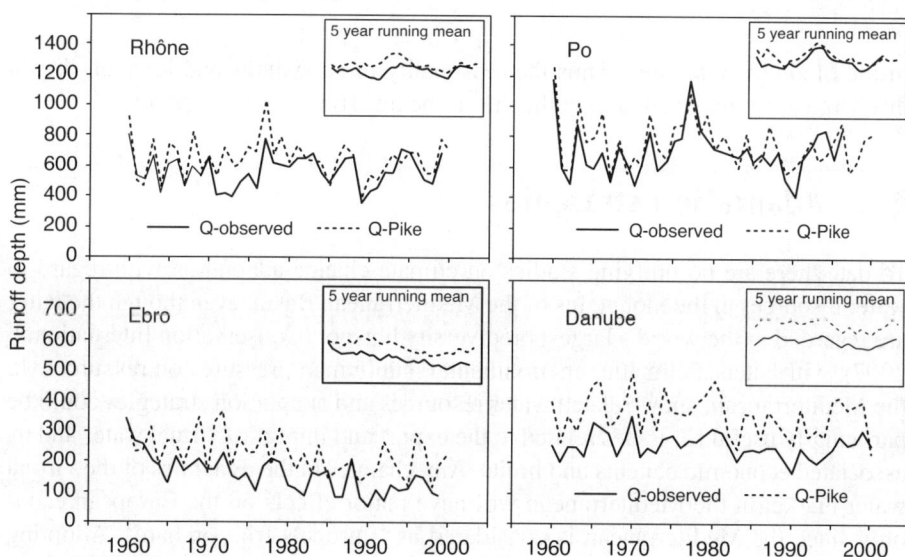

Figure 5.8   Comparison of observed and predicted (Q-pike) run-off depths for the Rhône, Ebro,
Po and Danube. Reproduced from Ludwig *et al.* (2009), with permission

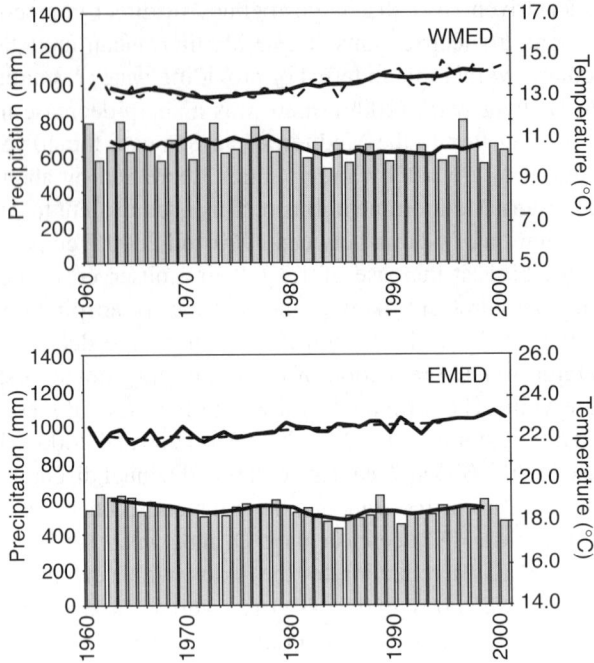

Figure 5.9 Changes in temperature (dashed lines), 5-year running mean of temperature (bold lines) and precipitation (grey bars) in the western Mediterranean (WMED) and eastern Mediterranean (EMED) basin between 1960 and 2000. Reproduced from Ludwig *et al.* (2009), with permission

in the Mediterranean area. Thus the seasonality of snowmelt- and ice-melt-driven discharge is an important aspect that has to be investigated in the future.

5.4 Adaptation strategies

To date there are no unifying studies on climate change adaptation with regard to water resources in the mountains of the Mediterranean Basin, even though the latter are regarded as the world's largest biodiversity hotspot (Conservation International, 2007), with intense climatic, environmental and human pressures on resources. In the Mediterranean, snow-related water resources and adaptation strategies could be particularly useful for issues related to the export and import of virtual water and its associated economic benefits and limits. Any changes in the dynamics of the virtual water market in the Mediterranean will have major effects on the European economy since the Mediterranean is considered as 'Europe's fruit orchard'. Adopting adaptation strategies for external climate and global forcing on mountain systems at an early enough stage, can ensure that Mediterranean agriculture is optimized and risks minimized. Compared to other regions worldwide with north-south relations, such as North and South America or Japan and southeast Asia, Europe has an important hinterland for its agricultural prosperity, the Maghreb region and Near East. Thus improved water management can have short- and long-term impacts linked to the regional, national and international scale.

The maintenance of a meteorological and environmental monitoring and measuring network in Mediterranean mountain regions could fill some important missing links for the World Meteorological Association (WMO), the United Nations Environment Programme (UNEP), the European Environment Agency and the Alpine Convention. The close links of these organizations to users and stakeholders should ensure the end-user-friendly proposition and monitoring of adaptation strategies. Within the framework of Alp-Water-Scarce, a project aiming to create an early warning system against water scarcity and to improve watershed management, some first steps have been made in this direction. Certain physical and socioeconomic factors in Mediterranean mountain regions could be optimized through decision-making processes, for example by optimizing water demand and availability, which influences water quantity and in turn water quality in downstream regions. By proposing the sharing of responsibilities between upstream and downstream water users, the availability and quality of water can also be improved. This is particularly true for the mountainous Mesta/Nestos basin at the Bulgarian-Greek boundary, the Meriç/Maritza river between Greece and Turkey, or the more complex situation of the Upper Jordan valley. Since snow seasonality already limits summer and autumn discharge, water strategies taking into account seasonal rationalization and groundwater replenishment need to be proposed for the future. In addition, due to steadily increasing mountain tourism, the rationalization of water consumption, during both summer and winter, is required particularly during the arid months and in geographically confined regions such as islands. Socioeconomic tools such as full recovery cost pricing in the water sector can contribute to the improvement of water resources management and optimal use of water storage.

Water- and resource-intensive technologies with high water losses by evaporation, such as the production of artificial snow for skiing, have to be subject to strict regulation. Conflicts between artificial snow production and drinking water demand are already occurring during low-discharge conditions in the winter months on a local scale in the Alps and Pyrenees and could rapidly escalate as a problem in other drier Mediterranean mountain zones. Also, the water quality of reservoirs storing water for snow production in mountain valleys is the cause of increasing concern as temperatures rise and good quality water sources become limited. Subsequently, drinking water collection zones can become contaminated by poor quality water derived from melting artificial snow. Since temperature is going to become more and more of a limiting factor for snow depth as well as artificial snow production, the energy and water consumption for snow making should be closely surveyed in the future.

Considering that climate change will impact the quantity and seasonal distribution of snow-derived discharge, dam reservoirs may more often function at their capacity limits in the future. Strategies proposing the changeover from pure energy to agricultural production will become necessary. In future, the regulation of minimal discharge in mountain rivers should not be restricted merely to energy demands but tuned to the requirements of the aquatic ecology and thus drinking water standards. Since most minimum river flows in the Mediterranean have been reduced by 70% primarily due to damming of mountain rivers, future strategies need to envisage a higher transmission of flow. Besides, more economic gains can be obtained from aquatic-oriented tourist activities such as fishing or rafting. Improvements in technology and lowering of desalination costs provide another tool in exceptional cases for increasing water availability, especially in very touristic semi-arid to arid regions. Minimal discharge should not be a mere statistical reflection of total flow quantity but should be adjusted to the requirements of the aquatic ecology, particularly in terms of water temperatures.

Land use and sustainable forest management is another issue of concern related to discharge quantity. In the eastern Mediterranean, a transition has occurred from forest to highly degraded forest patches. Should such forests be fully established again in the context of nature reserves and forest protection plans, the discharge in streams could be considerably reduced due to augmented evapotranspiration by forests.

In the Mediterranean, the tourist potential of mountain areas is poorly developed apart from winter sports. This is due to the competitive development of summer tourism in the Mediterranean coastal zones and the effects of globalization. Alternatives that are less water consuming should be established. However, the difference between ecotourism and hard mountain tourism such as off-road trucks and motorbikes and their varying impacts on the ecology should be considered. If the Mediterranean mountain zones are to be reinstated in value, suggestions are necessary for the sustainable valorization of physical and socioeconomic resources. Strategies for regulating livestock and grazing intensity as well as tourist intensity have to be established in order to improve the natural environment and biodiversity. There should be encouragement of knowledge transfer of adaptation strategies related to tourism and agriculture from other typical snow-dominated Mediterranean mountain regions both within the Mediterranean Basin and globally. Lessons could be learned from other snow-covered mountains such as the Sierra Nevada and the Rockies in California, the southern Andes in Chile, and to some extent also the Cape mountains in South Africa. All these mountain regions are particularly vulnerable to climate change impacts since, apart from the hydrologically limiting Mediterranean climate they are subject to comparably strong socioeconomic pressures.

A wider, more transdisciplinary approach to adaptation strategies is required for the future considering water use, agriculture, hydroelectricity, artificial snow production and four-season tourism.

5.5 Conclusion

The Mediterranean mountain regions are characterized by strong seasonal and regional contrasts and are facing a major scientific challenge in terms of climate change, hydrological change, natural hazards (mainly floods and droughts), adequate all-year-round water provision and nature conservation. Major monitoring networks need to be set up to bridge the inequality between the eastern and western Mediterranean in terms of data, water management, early warning systems and adaptation. The role, significance and protection of water resources in the upstream parts of the catchments require consideration. Snowmelt discharge regimes should be analyzed closely and integrated into seasonal hydrological forecasting. The potential climate change impacts on the duration and quantity of snowfall on hydrological systems as well as the economy, in particular agriculture and tourism, have to be foreseen. Climate change over the last 40 years is already having major impacts on Mediterranean climate and hydrology, therefore future research needs to be coordinated in an interdisciplinary and international manner to optimize the identification of particularly vulnerable regions and provide a portfolio of adaptation and mitigation options.

References

References marked as bold are key references.

Akman, Y. and Daget, Ph. (1971) Quelques aspects synoptiques des climats de la Turquie. *Bulletin de la Société languedocienne de Géographie* 5:596–310.

Alp-Water-Scarce. An Alpine Space Interreg project on Water Management Strategies against Water Scarcity in the Alps, 2008–2010, Coordinated by Carmen de Jong at the Mountain Institute, University of Savoy (www.alpwaterscarce.eu).

Aspen Global Change Institute (2006) *Climate Change and Aspen: an Assessment of Impacts and Potential Response.* Report, 147 pp.

Bacchi, B. and Villi, V. (2005) Runoff and floods in the Alps: an overview. In: de Jong, C., Collins, D. and Ranzi, R. (eds), *Climate and Hydrology in Mountain Areas.* John Wiley & Sons, Ltd, pp. 217–220.

Bales, R., Molotch, N.P., Painter, T.H., Dettinger, M.D., Rice, R. and Dozier, J. (2006) Mountain hydrology of the western US. *Water Resources Research* 42: issue 8, W08432; doi:10.1029/2005WR004387.

Beniston, M. (2003) Climate change in mountain regions – review of possible impacts. *Climate Change* 59:5–31.

Böhm, R., Auer, I., Brunetti, M., Maugeri, M., Nanni, T. and Schöner, W. (2001) Regional variability in the European alps 1760–1998 from homogenized instrumental time series. *International Journal of Climatology* 21:1779–1801.

Boudhari, A., Hanich, L., Boulet, G., Duchemin, B., Berjamy, B. and Chehbouni, A. (2009) Evaluation of the snowmelt runoff model in the Moroccan High Atlas Mountains using two snow-cover estimates, *Hydrological Sciences* 54:1–20.

Bradley, R.S., Keimig, F., Dıaz, H.F. (2004) Projected temperature changes along the American cordillera and the planned GCOS network. *Geophysical Research Letters* 31:L16210.

Castella, E., Hákon Adalsteinsson, H., Brittain, J.E. *et al.* (2001) Macrobenthic invertebrate richness and composition along a latitudinal gradient of European glacier-fed streams. *Freshwater Biology* 46:1811–1831.

Cazori, F. and Dalla Fontana, G. (1996) Snowmelt modelling by combining air temperature and a distributed radiation index. *Journal of Hydrology* 181:169–187.

Chaponnière, A., Maisongrande, P., Duchin, B. *et al.* (2005) A combined high and low resolution approach for mapping snow covered areas in the Atlas mountains. *International Journal of Remote Sensing* 26:2755–2777.

CIESM (2006) Fluxes of small and medium-sized Mediterranean rivers: impact on coastal areas. CIESM Workshop Monographs No. 30.

Conservation International (2007) Biodiversity hotspots: Mediterranean basins. Available at: www.biodiversityhotspots.org.

Corripio, J. and de Jong, C. (2005) Mountain waters: climate and hydrological sensitivity. *Hydrology and Earth System Sciences* 8:1015–1089.

de Jong, C. (2007) River resilience and dams in mountain areas. Talsperren *in Europa – Aufgaben und Herausforderungen*, 14. Deutsches Talsperrensymposium 7. ICOLD European Club Dam Symposium, Freising.

de Jong, C. (2009) Savoy – balancing water demand and water supply under increasing climate change pressures (France). *Regional Climate Change and Adaptation. The Alps facing the challenge of changing water resources.* Report, N. 8, ISSN 1725-9177. Copenhagen: European Environment Agency (EEA), pp. 81–84.

de Jong, C. and Schoeneich, P. (2009) Changes in the Alpine environment. How will the Alpine environment be tomorrow and for what activities? Recherche alpine: spécificité et devenir. *Journal of Alpine Research* 96:65–76.

de Jong, C., Machauer, C., Reichert, B., Cappy, S., Viger, R. and Leavesley, G. (2004) An integrated geomorphological and hydrogeological MMS modeling framework for a semi-arid mountain basin in the High Atlas, southern Morocco. In: Pahl-Wostl, C., Schmidt, S., Rizzoli, A.E. and Jakeman, A.J. (eds), *Complexity and Integrated Resources Management*, Transactions of the 2nd Biennial Meeting of the International Environmental Modelling and Software Society, iEMSs, Manno, Switzerland, 2004. ISBN 88-900787-1-5, pp. 736–741.

de Jong, C., Collins, D. and Ranzi, R. (2005a) *Climate and Hydrology in Mountain Areas*. John Wiley & Sons, Ltd.

de Jong, C, Whelan, F. and Messerli, B. (2005b) Mountain Hydrology. Special Issue. *Hydrological Processes* 19:2323–2449.

de Jong, C., Machauer, R., Leavesley, G., Cappy, S., Poete, P. and Schulz, O. (2005c) Integrated hydrological modelling concepts for a peripheral mountainous semi-arid basin in southern Morocco. In: Escadafal, R. and Paracchini, M.L. (eds), EU proceedings *Geomatics for Land and Water Management: Achievements and Challenges in the Euromed Context.* Luxembourg: Ispra, European Commission, JRC, pp. 219–227.

de Jong, C., Makroum, K. and Leavesley, G. (2006) Developing an oasis-based irrigation management tool for a large semi-arid mountainous catchment in Morocco. In: Voinov, A., Jakeman, A.J. and Rizzoli, A. (eds), *Proceedings of the iEMMs 3rd Biennial Meeting, "Summit on Environmental Modelling and Software"*. Burlington, USA: International Environmental Modelling and Software Society, pp. 6.

de Jong, C., Cappy, S., Finckh, M. and Funk, D. (2008) A transdisciplinary analysis of water problems in the mountainous karst areas of Morocco, Engineering and environmental problems in karst. *Engineering Geology* 99; 228–238.

de Jong, C, Lawler, D. and Essery, R. (2009a) Mountain hydroclimatology and snow seasonality – perspectives on climate impacts, snow seasonality and hydrological change in mountain environments. Special Issue of Mountain Hydroclimatology and Snow Seasonality. *Hydrological Processes* 23:955–961.

de Jong, C., Lawler, D. and Essery, R. (eds) (2009b) Special Issue of Mountain Hydroclimatology and Snow Seasonality. *Hydrological Processes* 23:955–103.

Diaz, H.F. (ed.) (2003) *Climate Variability and Change in High Elevation Regions: Past, Present & Future*. Dordrecht: Kluwer Academic Publishers, 282 pp.

Elsasser, H. and Messerli, P. (2001) The vulnerability of the snow industry in the Swiss Alps. *Mountain Research and Development* 21:335–339.

European Environment Agency (2005) Vulnerability and adaptation to climate change in Europe. EEA Technical Report No. 7, 84 pp. Copenhagen: EEA.

European Environment Agency (2009) Regional climate change and adaptation. The Alps facing the challenge of changing water resources. Report no. 8, 148 pp. Copenhagen: EEA.

FAO (2004) Watershed management case study: Mediterranean. Watershed management: a key component of rural development in the Mediterranean region. *Watershed Management and Sustainable Mountain Development*. Working Paper 4, 44 pp. UN Food and Agriculture Organization.

Ghosn, D., Vogiatzakis, I.N., Kazakis, G. *et al.* (2010) Ecological changes in the highest temporary pond of western Crete (Greece): past, present and future. *Hydrobiologia* 648:3–18.

García-Santos, G., Marzol, M.V. and Aschan, G. (2005) Water dynamics in a laurel montane cloud forest in the Garajonay National Park (Canary Islands, Spain). *Hydrology and Earth System Sciences* 8:1065–1075.

Gürer, İ. and Türksoy, M. (1981) Areal distribution of December–April period monthly precipitation totals of watershed of Seyhan Dam – selection of representative precipitation stations, I. National Meteorology Congress, Istanbul Technical University, İstanbul, Turkey, pp. 570–578 [original in Turkish].

Hoekstra, A.Y. (ed.) (2003) Virtual water trade. In: Proceedings of the International Expert Meeting on Virtual Water Trade. Value of Water Research Report Series No. 12, IHE Delft, pp. 244.

Jeftic, L., Milliman, J.D. and Sestini, G. (1992) *Climate Change and the Mediterranean: Environmental and Societal Impacts of Climatic Change and Sea level Rise in The Mediterrenean region*. Edward Arnold, pp. 673.

Latron, J., Soler, M., Llorens, P. and Gallart, F. (2008) Spatial and temporal variability of the hydrological response in a small Mediterranean research catchment (Vallcebre, Eastern Pyrenees). *Hydrological Processes* 22:775–787.

Latron, J., Soler, M., Llorens, P. and Gallart, F. (2009) The hydrology of Mediterranean mountain areas. *Geography Compass* 3:2045–2064.

Leavesley, G.H., Markstrom, S.I., Brewer, M.S. and Viger, R.J. (1996) The modular modeling system (MMS) – the physical process modeling component of a database centered decision support system for water and power management: *Water, Air and Soil Pollution* 90:303–311.

López-Moreno, J.I. and Garcia-Ruiz, J.M. (2004) Influence of snow accumulation and snowmelt on streamflow in the central Spanish Pyrenees. *Hydrological Sciences* 49:787–802.

López-Moreno, J.I., Beniston, M. and García-Ruiz, J.M. (2008) Environmental change and water management in the Pyrenees: Facts and future perspectives for Mediterranean mountains. *Global and Planetary Change* 61:300–312.

Ludwig, W., Meybeck, M. and Abousamra, F. (2003) Riverine transport of water, sediments and pollutants to the Mediterranean Sea. UNEP MAP Technical Report Series 141. Athens: UNEP/MAP.

Ludwig, W., Dumont, E., Meybeck, M. and Heussner, S. (2009) River discharges of water and nutrients to the Mediterranean and Black Sea: Major drivers for ecosystem changes during past and future decades? *Progress in Oceanography* **80:199–217.**

Margat, J. and Treyer, S. (2004) L'eau des Méditerranéens: situation et perspectives. Plan bleu. MAP *Technical Report Series* n° 158. Athens: UNEP; Plan d'action pour la Méditerranée; (Pam). www.planbleu.org/publications/mts158.pdf

McNeill, J.R. (1992) *The Mountains of the Mediterranean World: an Environmental History*. Cambridge University Press.

Menziani, M., Pugnaghi, S., Vincenzi, S. and Santangelo, R. (2005) Water balance in surface soil: analytical solutions of flow equations and measurements in the Alpine Toce Valley. In: de Jong, C., Collins, D. and Ranzi, R. (eds) *Climate and Hydrology in Mountain Areas*. John Wiley & Sons, Ltd, pp. 85–100.

Messerli, B. and Ives, J. (1997) *Mountains of the World. A contribution to chapter 13 of Agenda 21*. Parthenon Publishing Co.

Mitchley, J., Price, M. and Tzanopoulis, T. (2006) Integrated futures for Europe's mountain regions. Reconciling biodiversity, conservation and human livelihoods. *Journal of Mountain Science* 3:276–286.

MountSnowMed (2007) *Climate Change Impacts on Snow- and Ice-derived Water Resources and Adaptation Strategies in Mediterranean Mountain Regions*. Project proposal submitted to the 7th Framework Programme under the coordination of Carmen de Jong. Available from: http://www.hymex.org/?page=public/workshops/1/show&abs=public/workshops/1/abstracts/abstract_51.txt.

OECD: Agrawala, S. (ed.) (2007) *Climate Change in the European Alps. Adapting Winter Tourism and Natural Hazards Management*. OECD.

Ozenda, P. (1975) Sur les étages de végétation dans les montagnes du bassin méditerranéen. *Document de Cartographie Ecologique* 16:1–32.

Peyre, C. (1979) Recherches sur l'étagement de la vegetation dans le massif du Bou Iblane (Moyen Atlas oriental, Maroc). PhD thesis, University of Aix-Marseille, pp. 149.

Quézel, P. (1976) Les forêts du pourtour méditerranéen: écologie, conservation et aménagement. UNESCO. *Technical Note*, MAB 2:9–33.

Rhanem, M. (2008) Quelques aspects topoclimatiques de l'étagement de la végétation spontanée en montagne méditerranéenne, avec référence aux montagnes du Moyen et HautAtlas, (Maroc). *Quaderni di Botanica Ambientale e Applicata* 19:183–201.

Ricciardi, A. and Rasmussen, J.B. (1999) Extinction rates of North American freshwater fauna. *Conservation Biology* 13:1220–1222.

Rimmer, A. and Salingar, Y. (2006) Modelling precipitation-streamflow processes in karst basin: The case of the Jordan River sources, Israel, *Journal of Hydrology* 331:524–542.

Rivas-Martínez, S., Penas A. & Díaz, T.E. (2001) *Biogeographic Map of Europe: scale 1:16 mill.* Cartographic Service, University of León.

SAEFL: Greminger, P. (ed.) (2003) *Mountain Watershed Management. Lessons from the Past – Lessons for the Future*. Proceedings Environmental Documentation No. 165: Forests. Bern: Swiss Agency for the Environment, Forests and Landscape, FAO.

Schreier, H. (2003) Mountain watershed management: scaling up and scaling out. In: Greminger, P. (ed.) *Mountain Watershed Management. Lessons from the Past – Lessons for the Future*. Proceedings Environmental Documentation No. 165. Bern: Swiss Agency for the Environment, Forests and Landscape, pp. 29–42.

Schulz, O. and de Jong, C. (2004) Snowmelt and sublimation: field experiments and modelling in the High Atlas Mountains of Morocco. *Hydrology and Earth System Sciences* 8:1076–1089.

Shaban, A., Faour, G., Khawlie, M. and Abdallah, C. (2004) Remote sensing application to estimate the volume of water in the form of snow on Mount Lebanon. *Hydrological Sciences* 49:6–12.

Shaban, A., Khawlie, M., Abdallah, C. and Awad, M. (2005) Hydrological and watershed characteristics of the El-Kabir River, North Lebanon. *Lakes and Reservoirs: Research and Management* 10:93–101.

Figure 3.6 Late Pleistocene moraines in the Durmitor massif, Montenegro. These moraines formed during the Younger Dryas (12.9–11.7 ka) (photo by the author)

Figure 3.7 Vlasian Stage (190–127 ka) moraines in the Vourtapa valley on Mount Tymphi in northwestern Greece (photo by the author)

Figure 4.3 A small glacial laguna with permanent water during the summer season (photo by the authors)

Figure 4.7 Mount Etna. Astronaut photograph ISS013-E-62714 was acquired on 2 August 2006 with a Kodak 760C digital camera using an 800 mm lens, and is provided by the ISS Crew Earth Observations experiment and the Image Science & Analysis Group, Johnson Space Center (http://earthobservatory.nasa.gov/)

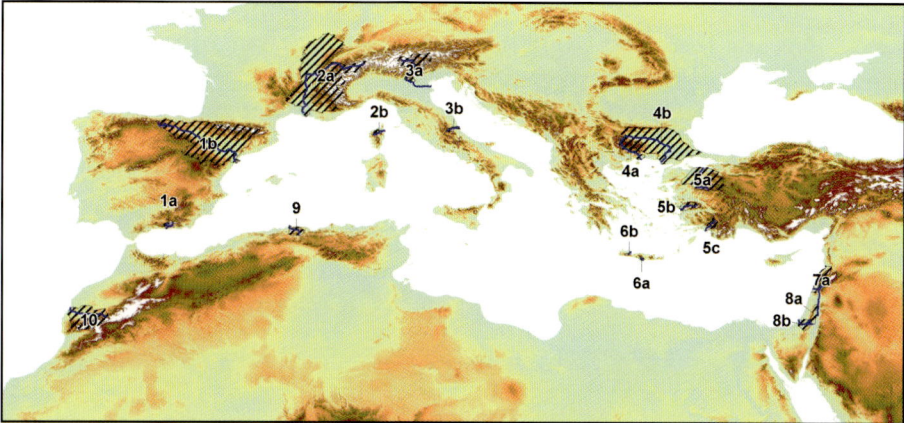

Figure 5.1 Mediterranean mountain river basins (including mountain ranges on mainland and islands) of pilot sites corresponding to Table 5.2

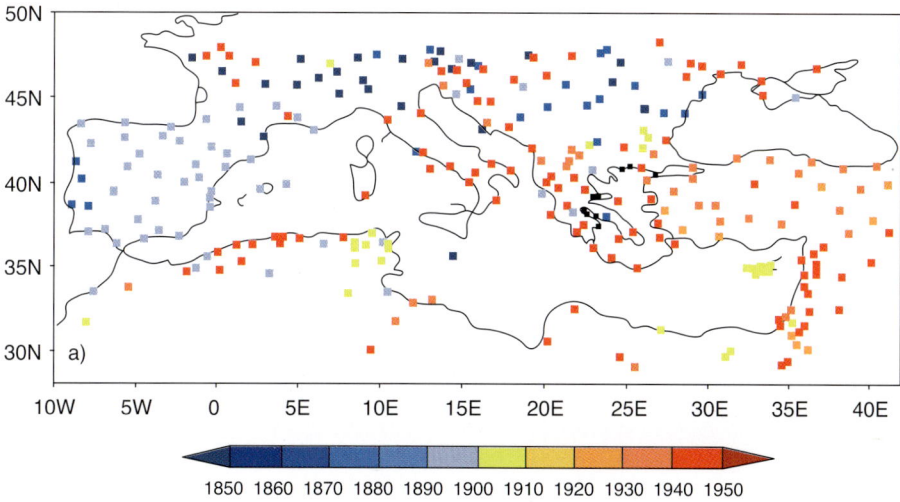

Figure 5.2 Dates of beginning of precipitation records (in decades) across Mediterranean. Reproduced from Xoplaki *et al*. (2004), with permission

Figure 5.5 Regions with artificial snow production in the Mediterranean mountain regions. Size of symbol corresponds roughly to amount of snow production closely related to relative water volume abstracted. Reproduced from Carmen de Jong (2007), with permission

Figure 6.4 Dolines on the central Lefka Ori massif, Crete (photo by the author)

Figure 7.2 (a) View of Mount Tabor (source: Wikimedia). (b) Theophanes the Greek, Transfiguration, fourteenth century (source: Wikimedia) (© Tretyakov Gallery, Moscow)

Figure 7.6 (a) Modern cottages in Pertouli, and (b) church domes overlooking caldera in Oia, Santorini (photographs by Theano S. Terkenli)

Figure 8.3 Abandoned old terraces in the White mountains of Crete colonized by phryganic vegetation (photo by the author)

Figure 8.4 A landscape denuded by bauxite mining on Giona Mountain, central Greece (photo by the author)

Figure 9.2 Mountainous areas analyzed in this chapter. Mountain areas have been defined following the UNEP-WCMC classification. Analyses of climatic and Normalized Difference Vegetation Index (NDVI) trends are just within the strict limits of the mountain ranges in this figure. Pyrenees (pink), Apennines (orange), Dinaric Alps (green), Pindos (blue), Taurus (purple) and Atlas (red)

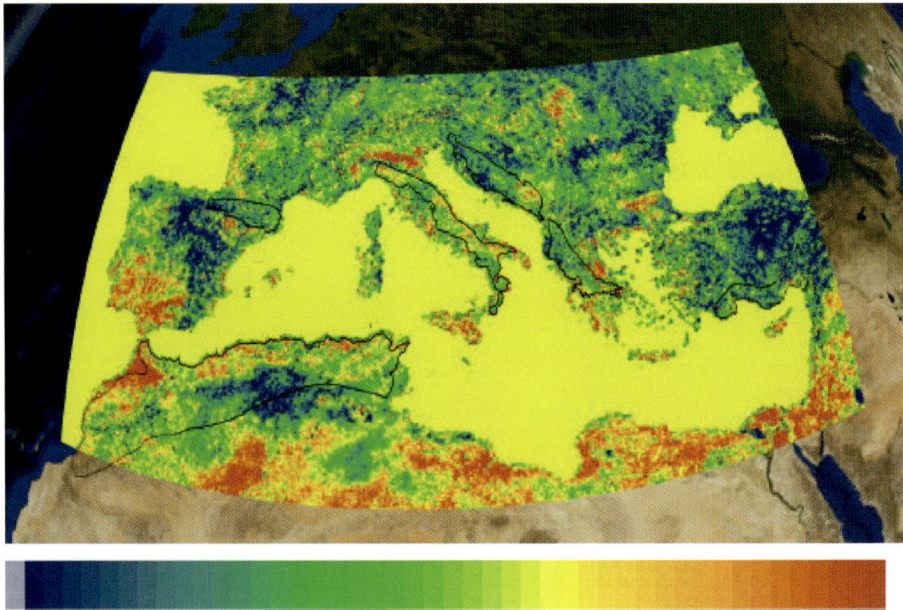

Positive NDVI trends　　　　　　　　　　　　　　　　　**Negative NDVI trends**

Figure 9.4 Trends in vegetation activity in the 25 years from 1982 to 2006. Values of the Normalized Difference Vegetation Index (NDVI) were obtained from the Global Inventory Modelling and Mapping Studies (GIMMS)

Skoulikidis, N.Th. (2009) The environmental state of rivers in the Balkans – A review within the DPSIR framework. *Science of the Total Environment* 407:2501–2516.

Strasser, U. and Etchevers, P. (2005) Simulation of daily discharges for the Upper Durance catchment (French Alps) using subgrid parameterisation for topography and a forest canopy climate model. *Hydrological Processes* 19:2361–2374.

Tonkov, S., Panovska, H., Possnert, G. and Bozilova, E. (2002) The Holocene vegetation history of Northern Pirin Mountain, southwestern Bulgaria: pollen analysis and radiocarbon dating of a core from Lake Ribno Banderishko. *The Holocene* 12:201–210.

Villa, S., Negrelli, C., Maggi, V., Finizio, A. and Vighi, M. (2006) Analysis of a firn core for assessing POP seasonal accumulation on an Alpine glacier. *Ecotoxicology and Environmental Safety* 63:17–24.

Viviroli, D., Weingartner, R. and Messerli, B. (2003) Assessing the hydrological significance of the world's mountains. *Mountain Research and Devlopment* 23:32–40.

Viviroli, D., Dürr, H.H., Messerli, B., Meybeck, M. and Weingartner, R. (2007) Mountains of the world –water towers for humanity: typology, mapping and global significance. *Water Resources Research* doi:10.1029/2006WR005653.

Watson, R.T., Zinyowera, M., Moss, R.H. and Dokken, D. (1997) The regional impacts of climate change: an assessment of vulnerability. Special Report. Intergovernmental Panel on Climate Change.

Xoplaki, E., Gonzalez-Rouco, J.F., Luterbacher, J. and Wanner, H. (2004) Wet season Mediterranean precipitation variability: influence of large-scale dynamics and trends, *Climate Dynamics* 23:63–78.

6
Biogeography

Ioannis N. Vogiatzakis

6.1 Introduction to mountain biogeography

Mountains are hotspots of biodiversity worldwide (Myers *et al.*, 2000; Körner and Spehn, 2002). They support a large number of endemic species, as well as being centres of origin of important crops (Hamilton and McMillan, 2004). Due to their altitudinal range, mountains exhibit over a few kilometres of vertical distance climatic regimes that can be observed along longitudinal or latitudinal gradients in lowland areas. Therefore they support a wide range of vegetation formations demonstrating a distinct zonation. In addition to the diverse vegetation resulting from steep environmental gradients, there are also many ecotones between these zones. The direct influence of orogenesis on biogeography is twofold: it has created a mosaic of habitats and acted as a barrier to migration. Additional habitat-creating factors include landforms, hydrology and soils. The terrain in mountain environments is highly fragmented and diverse topographically. Mountain hydrology also contributes to beta diversity due to the presence of glaciers, seasonal snow cover, ice and seasonally frozen ground (Nagy and Grabherr, 2009).

Mountains are often located on the borders of different biogeographical regions, and the Mediterranean mountains are no exception. Located at the crossroads of three continents their flora comprises different phytogeographical elements ranging from Euro-Siberian to Arctic-Alpine, to Irano-Turanian in the eastern Mediterranean, as particularly demonstrated in the mountain flora of Crete, Cyprus and Turkey (Table 6.1). Mediterranean mountain floras share common characteristics in origin, lifeform and morphology as well as common genera and species. However, local species composition is related to biogeography, the length and degree of isolation and speciation. In the Mediterranean the north–south floristic affinities are stronger than those of east–west. This is also reflected in the dominant species in the altitudinal belts of the region (Quézel and Médail, 2003). The Mediterranean mountains host many regional and local endemic species, some of which are relicts of past biogeographical patterns (Médail and Verlaque, 1997). In particularly the presence of Tertiary relictual vegetation is evident in the mountains of

Mediterranean Mountain Environments, First Edition. Edited by Ioannis N. Vogiatzakis.
© 2012 John Wiley & Sons, Ltd. Published 2012 by John Wiley & Sons, Ltd.

Table 6.1 Percentage frequencies of different phytogeographical elements in the mountain flora of Greece

	Crete	Peloponessos	Sterea Ellas	S. Pindos	N. Pindos	E. Central	N. Central	Northeast
Cosmopolitan	14.3	11.5	14.8	17.1	19.4	16.9	20.2	21.8
Central and S. European	7.8	13.5	16.1	18.7	21	17.3	21	22
Balkanic and Anatolian	12.9	13	9.9	8.4	8.2	12.9	8.8	12
Mediterranean	14.7	15.2	13	13.8	11.1	19.8	10.9	10.8
Balkanic and Italian	2.8	4.4	4.9	7.1	6	4	4.8	2.9
Balkan endemic	2.3	11.5	14.9	20.2	21.9	18.3	22.4	20.2
Greek endemic	6.5	16.3	15.4	9.5	5.6	5.8	4.5	2.4
Single area endemic	24.4	5.7	3.6	0.3	1.4	0	0.1	0.8
Single mountain endemic	11.1	4.4	2.5	0.2	1.1	0	3.7	3.2
Other	3.2	4.5	4.9	4.7	4.3	5	3.6	3.9
Total number of species*	217	540	751	609	780	278	868	665

Reproduced from Strid (1993).
*Absolute number.

Morocco, Greece, Lebanon and southwest and southeast Turkey (Quézel and Médail, 2003). In a recent study, Médail and Diadema (2009) identified 33 mountainous areas within 52 refugia in the Mediterranean Basin. Some of these mountains had already been identified as regional biodiversity hotspots (Médail and Quézel, 1997) and global centres of plant diversity (Table 6.2) (Davis *et al.*, 1994)

Geology is another factor that contributes to biodiversity richness. The association of lithologies with plants is one of the major factors determining plant and community distribution patterns in mountain environments. The diversity of parent material gives rise to a rich flora but also to specialized plants as demonstrated by the limestone flora of Mount Olympus in Greece and the Taurus Mountains of Turkey, or the serpentinophilous flora of Troodos in Cyprus and the siliceous flora of the Baetic Ranges in Spain (Mota *et al.*, 2002).

Although species richness follows the general rule whereby increases in altitude are characterized by a decrease in richness, this relationship is not monotonic and there are rich pockets of habitats. Another rule that applies in mountain areas, that of the increase in endemism as the altitude increases, is a consequence of past isolation and resulting speciation. In biogeography, mountain tops have often been considered as similar/analogous to islands since they provided refuge to arctic-alpine species during the interglacial periods: 'Terrestrial or continental habitat islands (MacArthur & Wilson) are these relatively small areas of land or water surrounded by, and isolated from, larger ecologically similar areas by an extensive area of dissimilar habitat' (Tivy, 1993, p. 258).

Plants become smaller as the altitude increases, a result of harsher conditions at higher zones. The main limiting factors, typical of mountain environments, are low

Table 6.2   Mountain centres of plant diversity (CPD) in the Mediterranean

Location	Plant species	Threats
Baetic and Sub-Baetic Mountains, Spain	c. 3000 plants	Forest fires, pollution, tourist development
Gudar and Javalambre massifs Spain	c. 1500 plants	Tourism development
Pyrenees	c. 3000 plants, 200 endemic	Tourism, erosion
Southern and central Greek mountains	c. 4000 plants	Fire, grazing
Crete*	c. 1600 plants, 10% endemic*	Fire, tourism, agriculture
Troodos, Cyprus	1650 plants, 62 endemic	Fire, road-building
Levantine uplands	c. 4160 plants, c. 635	Logging, agriculture, urbanization
High Atlas, Morocco	c. 1000 plants: 160 endemics confined to the high mountain zone	Population pressure, overexploitation of resources
Isaurian, Lycaonian and Cilician Taurus, Turkey	c. 2500 plants, 235 endemic	Agriculture, tourism, exotic species

Based on data from Davis *et al.* (1994).
*CPD and figures include the mountain massifs of the island

temperatures, temperature extremes, prevalence of winds and the very short growing season (Tivy, 1993). The main functional types in these zones include tussock-forming grasses, low-stature shrubs, mat-forming graminoids, legumes with N-fixing symbionts, and rosette-forming, non-legume herbs (Nagy *et al.*, 2003). Invertebrates exhibit similar patterns, that is they decrease in number as well as body mass as the altitude increases. In recent years mountain chains have been places of discovery of both plants – such as *Horistrisea dolinicola* (Egli, 1991) in the Psiloritis mountain of Crete – and invertebrates – such as the carabid beetle *Relictocarabus meurguesae* in the High Atlas of Morocco (Blondel *et al.*, 2010). The predominant plant lifeforms, hemicryptophytes and chamaephytes, reflect the harsh environmental conditions dominating in these altitudes.

At the end of the Pliocene, Mediterranean mountain vegetation showed similarities to that of the present day. According to Pignatti (1978), in the largest mountains surrounding the Mediterranean (Alps, Pyrenees, Atlas), deciduous forests dominated by beeches and oaks were replaced by evergreen forest at higher altitudes; open coniferous communities (with pines, firs and cedars) were also present, whereas spiny shrub formations dominated the higher mountain slopes and summits (for a discussion of the Quaternary see Chapter 2). Today the Mediterranean is characterized by vegetation types such as forests, open woodlands, maquis, garrigue, phrygana and steppe. Most of these categories occur in an altitudinal zonation, following a gradient from the lowlands to the alpine zone that demonstrates a decrease in structural complexity of vegetation cover. In addition there are many azonal habitats – products of the geology, hydrology and their interaction that host a variety of important communities and species.

6.2 Vegetation

The foothills of Mediterranean mountains are covered with what is widely perceived as Mediterranean vegetation, mostly pre- or post-forest formations with an arborescent cover and an evergreen stratum, comprising such genera as *Arbutus*, *Erica* and *Pistacia* (i.e. arborescent matorral). Above 1000 m the deciduous and evergreen forests are dominated by oak species such as *Quercus cerris, Q. ilex, Q. coccifera* and *Q. suber* while other elements include *Acer* spp., such as *Acer sempervirens, A. hyrcanum, A. monspessulanum* and *A. obtusifolium*, but also genera unique to the mountains of the Mediterranean such as *Zelkova* and *Berberis*.

Mountainous forests, particularly from 1200 m to 1600 m, are dominated by coniferous species (Quézel, 2004). Pines are the most widespread conifers, including *Pinus nigra, P. halepensis, P. brutia* and *P. pinaster* in the western Mediterranean. Among pines *P. brutia* in particular dominates in the east Mediterranean, especially in the mountains of Lebanon, southern and western Turkey, Cyprus and Crete. However, there are other important coniferous species, including cedars in the Atlas Mountains, Cyprus, Lebanon and Turkey, junipers in the Spanish Sierras and the Atlas Mountains, and firs in Turkey, Lebanon and Spain.

The most widely used altitudinal zonation for Mediterranean vegetation (Figure 6.1), proposed by Quézel (1981a), is based on latitude, and corresponds to

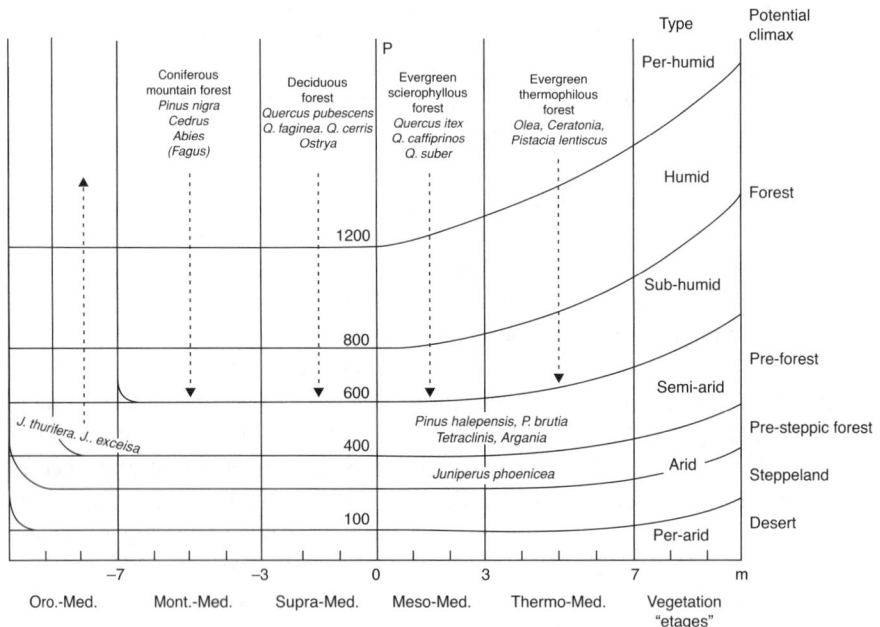

Figure 6.1 Schematic representation of the major types of vegetation structure around the Mediterranean according to the bioclimatic types and 'etages' of vegetation. p, annual mean rainfall; m, mean precipitation of the coolest month. Only a few species are indicated. Reproduced from Quezel (2004), with permission

climatic variations, especially temperature. Naturally there are many variations between mountains (due to geology, altitude and human history), with some mountains supporting unique communities/assemblages. The (upper) Montane Mediterranean zone is dominated either by beech or conifers, including pines, cedars and firs. In the Taurus mountains between 1200 and 2000 m, *Pinus nigra* ssp. *palassiana* forests are found, which are often associated with *Cedrus libani* and *Abies cilicica* (Kaya and Raynal, 2001). *Cedrus libani* forests are confined to the highest zones of the eastern Mediterranean, reaching about 2200 m. Some of the best examples are seen in the Lebanon Mountains, while less extensive and often relict stands can be found in Syria and the Amanus and Taurus Mountains. Above 1000 m in the Troodos mountains of Cyprus there are stands of the endemic Cyprus cedar *Cedrus libani* ssp. *brevifolia* as well as *Juniperus foetidissima* above 1600 m (Tsintides *et al.*, 2002). *Juniperus excelsa* and *J. foetidissima* forests are also common in the Taurus Mountains particularly in places where cedar and pine forest have been degraded (Kaya and Raynal, 2001). In the High Atlas of Morocco there are *Cedrus atlantica* relict stands, while in the mountains of Crete and Sicily the endemics *Zelkova abelicea* and *Zelkova sicula* are respectively present. In the Levantine uplands *Quercus calliprinos* forest and scrub, *Abies cilicica* forest and *Cedrus libani* forest occur. In Corsica a sharp north–south gradient in terms of plant communities typifies the highest elevations of this zone. *Pinus laricio* dominates on south-facing slopes while silver fir (Abies alba) and beech (*Fagus sylvatica*) are dominant on the north-facing slopes (Mouillot *et al.*, 2008).

Nagy and Grabherr (2009) discuss the difficulty in delineating the treeline in the European Mediterranean due to the absence of trees and their replacement by thorny shrubs. The treeline-forming trees in the Mediterranean mountains include *Pinus heldreichii* in the Hellenids, *Cupressus sempervirens* in Crete, *Juniperus communis* ssp. *hemisphaerica* in the Sierra Nevada and *Juniperus excelsa* in the Atlas Mountains (Figure 6.2). In Greece the timberline lies at c. 1800 m, a criterion used to delineate the country's mountain flora (Strid, 1986). However, in the highest Greek mountain, Olympus, the treeline reaches 2300 m. In the High Atlas the treeline lies at approximately 2400 m, formed by *Juniperus excelsa* and *Juniperus foetidissima*, while in the Hellenids this role is played by *Pinus heldreichii* and in the Taurus by *Juniperus thurifera* (Nagy and Grabherr 2009). In the Lefka Ori of Crete the upper limit of forest growth on the southern side of the massif is at 1600–1650 m, while on the northern side the limit is up to 150 m higher. This limit is significantly lower compared to more extensive mountain ranges such as the Alps, where the limit rises to 2400 m. (Turland *et al.*, 1993).

The zones above the treeline are rich in perennial but poor in annual species. The alti-Mediterranean zone usually includes dwarf junipers mixed with diverse grass communities of *Bromus, Festuca, Poa, Phleum* and other perennial species (Blondel *et al.*, 2010). In Corsica, for example, the high summits are characterized by the dominance of dwarf shrub alder, juniper and maple (Moullot *et al.*, 2008) whereas the corresponding zone in the Sierra Nevada has open dry grassland with endemics such as *Eryngium glaciale* and *Festuca clementei*. The main vegetation formations covering the stony slopes of the highest mountain zones in the

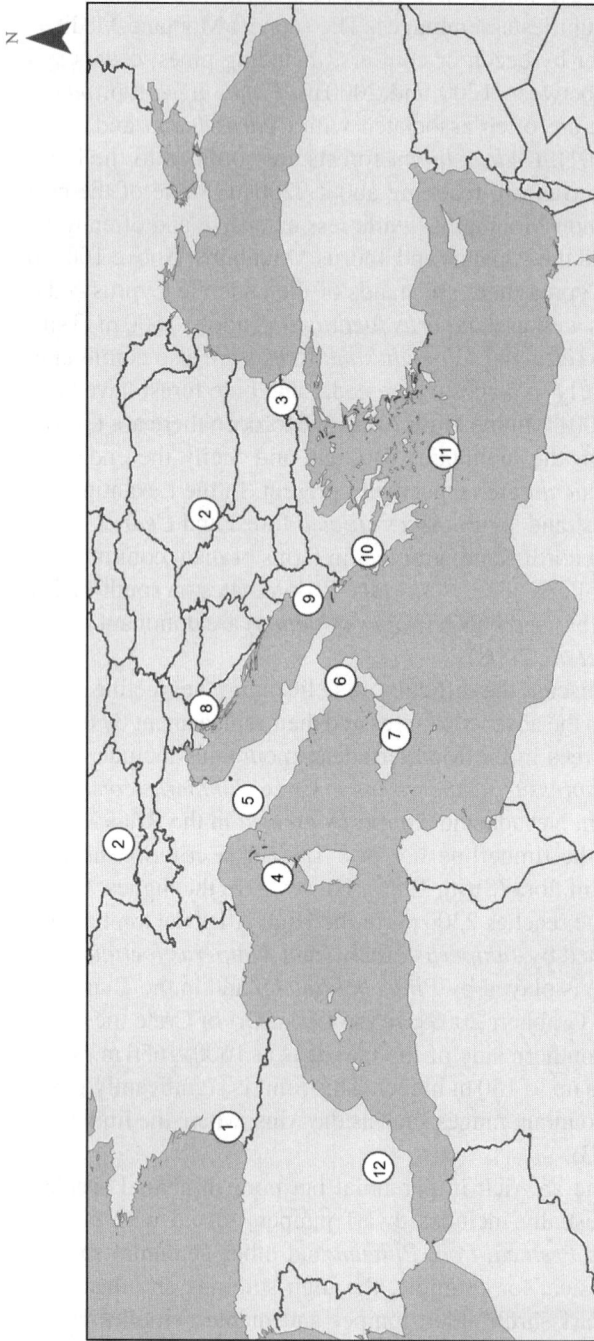

Species and mountain areas:

1. Pinus uncinata (Pyrenees)
2. Larix decidua, Picea abies, Pinus cembra, P. mugo, Alnus viridis (Alps, Carpathians)
3. Picea abies, Pinus mugo (Rila)
4. Alnus viridis ssp. suaveolens (Corsican mountains)
5. Fagus sylvatica, Pinus mugo (central Apennines)
6. Pinus leucodermis, Juniperus communis ssp. hemisphaerica (southern Apennines)
7. Genista aetnensis [Abies nebrodensis] (Mt Etna)
8. Pinus mugo (northern Dinarids)
9. Pinus peuce, P. heldreichii (southern Dinarids)
10. Pinus heldreichii (Hellenids)
11. Cupressus sempervirens (Crete)
12. Juniperus communis ssp. hemisphaerica, J. sabina, Genista baetica (Sierra Nevada)

SCALE 1:20.000.000

Figure 6.2 Treeline species in Mediterranean mountains. Reproduced from Nagy and Grabherr (2009), with permission

Figure 6.3   Oro-Mediterranean formations in the northwestern Lefka Ori (photo by the author)

Mediterranean (oro-Mediterranean and alti-Mediterranean zones) are prickly scrub or spiny cushion-shaped dwarf shrubs (Figure 6.3). Dwarf shrublands are known to occur in the Mediterranean and the Irano-Turanian regions, as well as the South American Andes. These formations, also known as hedgehog heath or cushion heath shrubs, are one of the most characteristic vegetation types of the mountains of Italy, Spain, Greece and northern Africa (Quézel, 1981c; White, 1983). They represent a climax formation that reflects the environmental conditions. The term used to describe these formations is *pelouse écorchée* (Zaffran, 1990). They have three characteristics:

- dominance of spiny chamaephytes in cushions;

- presence of bare ground between the tufted species;

- presence of a moderate pasture for sheep or wild goats.

In the Lefka Ori of Crete, the dominant species include *Berberis cretica* L. and *Euphorbia acanthothamnos,* with *Juniperus oxycedrus* L. ssp. *oxycedrus* or communities of spiny cushion-shaped dwarf shrubs such as *Acantholimon androsaceum* (Jaub. & Spach) Boiss., *Astragalus angustifolius* Lam. and *Satureja spinosa* L. (Vogiatzakis. 2000). In Morocco the zone corresponds to the climatic climax between the treeline and 3900 m and includes species such as *Alyssum spinosum, Amelanchier ovalis, Berberis hispanica* and *Prunus prostrata* (White, 1983). Above the alti-Mediterranean zone an additional zone may be present; this cryo-Mediterranean zone is present only in the highest mountains of the Basin such as the Atlas Mountains. This zone is devoid of vegetation apart from a range of widespread alpine species in rocks and screes.

In addition to the formations described above, many mountain habitats and plant communities are independent of this elevation gradient but instead are associated with geomorphological or hydrological processes and features. Cliffs are a conspicuous feature of the Mediterranean Basin, whose presence is often associated with mountains. The main characteristics of cliffs are vertical surfaces, lack of a well-developed soil layer, severe drought effects and extreme diurnal temperatures, and their importance for the flora has long been recognized (Turland *et al.*, 1993). According to Snogerup (1971), there are no sharp limits between communities on a single cliff, and while the species composition in two adjacent cliffs can be entirely different the vegetation formation can be identical. Cliffs have provided habitats for some highly specialized plants, many of which lack the competitiveness to survive elsewhere. Those plants able to tolerate harsh cliff conditions are termed chasmophytes. Moreover, cliffs serve as refuges and may host a high number of relict endemic species, such as *Petromarula pinnata* in Crete, protecting them from grazing and environmental change.

Another feature that dominates the highest mountain summits are scree slopes consisting of angular blocks of various shapes and sizes, which provide habitats for many plants. Scree formation in hot and dry regions of the world is driven by the sudden expansion and contraction of the rock. Rupicolous chasmophytic vegetation is more diverse in calcareous than siliceous rocks on the High Atlas Mountains (Quézel, 1981c) while there is differentiation in community composition between fixed and mobile screes (Vogiatzakis and Griffiths, 2006). In the Lefka Ori of Crete, calcareous scree formations resulting from limestone weathering are also abundant above 1900 m. Some of the species adapted to life on screes are *Cicer incisum, Peucedanum alpinum* and the endemics *Silene variegata* and *Viola fragrans*.

In depressions where water and soil accumulate from snowmelt there are summer dry meadows (e.g. in the Pindos and Mount Olympus in Greece and in the High Atlas in Morocco), with species such as *Festuca, Trifolium* and *Plantago*, while the impermeable nature of siliceous bedrock gives rise to waterlogged pozzines or mires with distinct plant communities (Nagy and Grabherr, 2009). The Taurus Mountains of Turkey are a good example, with high plant community richness due to the variety of geomorphological and hydrological processes occurring. Some of these communities include the vegetation of hilltops and exposed ridges, snow-beds and meltwater communities and azonal hydrophytic units (Parolly, 2004). In the mountains of Crete, the Dinaric Alps and Spanish Sierras, the abundance of carbonate rocks has led to the development of karstic formations (Figure 6.4; see Chapter 4). Typical features include dolines, poljes, deep gorges and extensive cave systems. In the Lefka Ori massif of Crete there are numerous dolines, for example, mainly dominated by the evergreen shrub *Berberis cretica*. They support more perennial than annual plants and are rather poor in endemic species confined strictly to this habitat compared to scree slopes and mountain pastures (Egli, 1991). Waterlogging

Figure 6.4 Dolines on the central Lefka Ori massif, Crete (photo by the author). (*A full colour version of this figure appears in the colour plate section*)

may occur in dolines on compressed soils, restricting growth to a few specialized plants in the centre and more diverse vegetation confined to the doline edge. In dry dolines, the established vegetation, mainly comprising chamaephytes, hinders soil compression. The hydrological conditions are optimal for plant growth with high water uptake and storage capacity (Egli, 1989).

Plants growing in unfavourable environments have adapted to allow them to survive the harsh conditions, and this is also true for alpine plants. The main limiting factors in the alpine zone are temperature extremes, prevalence of strong winds and the very short growing season (Tivy, 1993). Therefore, many plants in alpine mountain zones are characterized by a low (cushion-like) growth form, which protects them from the strong cold winds and allows them to survive on the ground surface where the temperatures are higher. Alpine plants tend to be perennials rather than annuals, in order to cope with the limited growing season. For example, in the mountain areas of Greece this is the predominant lifeform followed by annual, woody/suffruticose taxa (Table 6.3). Extensive root systems, which facilitate nutrient uptake and water retention during droughts, are also common (Nagy and Grabherr, 2009). Apart from the harsh physical environment, grazing is one more 'obstacle' to plant survival in mountainous regions. Plant species adapt to grazing in many ways. Most of them have protective spines on their stems or leaves, or contain substances that make them unpalatable to grazing animals.

Table 6.3   Plant lifeforms (as a percentage of total) in the mountain areas of Greece

	Annual-biennial	Perennial	Woody-suffruticose	Geophytes	Grass-like	Spiny
Crete	17.8	56.2	14.7	8.8	7.8	5
Peloponissos	11.3	64.4	12	9.1	9.4	4.4
Sterea Ellas	11.7	64.6	11.3	8.5	9.5	3.6
S. Pindos	11.5	65	10.5	6.7	11.8	3.3
N. Pindos	10.8	65.6	10	6.9	11.8	2.7
E. Central	12.6	62.9	11.5	8.6	9.7	3.2
North Central	10.5	65.5	10.2	7.1	11.4	2.2
North East	9.3	64.4	13.8	7.5	10.2	2.2

Reproduced from Strid (1993).

6.3 Flora

Phytogeographically there are many elements other than Mediterranean ones (see Table 6.1) in the flora of the mountains in the Basin. For examples, the Pyrenees are in a transition area between Central and Mediterranean Europe and contain a small Mediterranean phytogeographical unit mainly in the south and east of the main axis of the Cordillera (Davis *et al.*, 1999). In Greece the Mediterranean element reaches its peak in the mountains of the Central East, followed by the massifs in the Peloponnese and Crete (Table 6.1). The Dinaric Alps contain a range of alpine continental and Mediterranean species but also species of boreal origin (Tvrtkoviç and Veen, 2006). The presence of a high number of endemic species in the Baetic ranges of Spain resulted in the area being recognized as a distinct biogeographical unit (Mota *et al.*, 2002). The Baetic and sub-Baetic Mountains share many species with the mountains of North Africa. Although the rule of thumb suggests that the flora of the Mediterranean mountains does not display high overall species richness (Medail and Verlaque, 1997) there are a few exceptions. The flora of the Rif mountains of north Africa exceeds 2000 species per $15\,000\,km^2$ (Moore *et al.*, 1998) whereas the Greek mountain flora (above 1500 m) comprises 1600 species (Strid, 1986; Strid and Tan, 1991).

Many mountain species in the Mediterranean have evolved from lowland species as a result of ecotypic differentiation on an altitudinal gradient combined with geographical and/or ecological differentiation (Strid, 1985). For example, the flora of the Cretan mountains comprises relict species, many of which are endemic derivatives of lowland species and species that also occur in the continental Greek mountains such as *Scutellaria hirta* from *Scutellaria sieberi* and *Bellis longifolia* from *Bellis sylvestris* (Greuter, 1972). A large number of lowland species reach the altitudinal zone between 1000 and 1500 m. Some species such as *Euphorbia acanthothamnos*, *Verbascum spinosum* and *Arum creticum* can survive in the high mountain tops above 2000 m without being modified.

Rising to heights of 2100 m, Mount Lebanon is a critical habitat for the Lebanese cedar in Lebanon. The species is not threatened at a global level, but only small patches remain in Lebanon. Outside Lebanon there are two major groups of *Cedrus atlantica* populations, one distributed through the Rif and Middle Atlas mountains in Morocco and the other through the Algerian Tell Atlas and Aurès mountains as well as the Middle Atlas (Terrab *et al.*, 2008).

In the Madonie mountains of Sicily there are relict forests of *Ilex aquifolium* and a high number of endemics and narrow endemics. The active volcano of Mountain Etna in the east of Sicily is also rich in endemics such as *Astragalus siculus*, *Genista aetnensis* and *Betula aetnensis*. The significant difference in the vegetation between the Madonie mountains and Etna reflect their vegetation history. Madonie has a well-developed Tertiary vegetation while Etna shows a post-glacial vegetation (Pignatti, 1978). The mountains of Sicily are the southern distribution limit for a number of northern and central European species such as beech (*Fagus sylvatica*), which is widely present in the Nebrodi mountains, and yew (*Taxus baccata*) (Benedetto and Giordano, 2008).

6.3.1 Endemism

Both insularity and mountain terrain are considered to be significant causes of high endemism. The analysis of data by Medail and Verlaque (1997) for comparable territories in the Mediterranean and throughout southern Europe showed that:

- mountain isolation has generally been more favourable to endemism than insularity;

- often the degree of endemism decreases as floristic richness increases;

- rates of endemism range from 10% to 42%.

Rates of endemism of over 20% occur in the Baetic-Rifan complex on either side of the Strait of Gibraltar, in the Middle Atlas and High Atlas in Morocco, in the Iberian Sistema Central, in the Pindos Mountains of Greece, in the southern mountains of Turkey (Taurus and Amanus) and the Lebanon mountain range (Médail and Quézel, 1997). According to Dominguez *et al.* (1996), 60% of the Iberian endemic flora occurs in high mountain habitats. For example, in the Pyrenees chain there are 180 endemics confined to the alpine zone (Gómez *et al.*, 2003), while the alpine communities of the Apennines contain more endemic and rare species than their eco-functional counterparts in the Alps (Pedrotti and Gafta, 2003)

Mountain endemism may be the result of specific and localized factors such as discrete orogenies and rare substrates (Kruckenberg and Rabinowitz, 1985; Major, 1988). Limestone, serpentine and gypsum are well known to botanists for being associated with this phenomenon. For example, in Greek mountains limestone and serpentine in particular host the largest concentration of endemics (Strid and

Papanikolaou, 1985), while in the Baetic ranges of Spain the centres of endemism are distinguished on the basis of the substrate (calcareous and siliceous) (Mota *et al.*, 2002). Despite the problems in comparing the endemism rate among different mountain ranges in Europe, Favarger (1972) concluded that the southern mountains have a higher percentage of endemism (30–40%) than the northern ones (12–18%). The author suggests this is probably due to:

- a north–south gradient reflecting more favourable climate conditions;

- a comparatively minor influence of glaciation in southern European mountains, providing refugia for species that have become extinct elsewhere;

- the high proportion of species that central European mountains have in common.

There are 405 endemic species in the Greek mountain flora (above 1500 m) (Strid, 1986; Strid and Tan, 1991). On Mount Olympus, for example, there are approximately 150 species above 2400 m, half of which are endemic to the Balkan peninsula with a dozen of those confined to Olympus (Strid, 1995).The flora of the Cretan mountains is considered to be poor in absolute number of species (217 taxa), compared to the remaining mountainous regions in Greece, as a result of the dry and harsh conditions. However, there are 132 plant species endemic either to Greece or the Cretan area occurring in the Lefka Ori massif. Strid (1995), discussing the phytogeographical elements in the mountain flora of Greece, emphasizes the importance of the Cretan mountains as refuges of endemism. Both regional and local endemism in Greece increase in a southerly direction, culminating in the Lefka Ori of western Crete, which displays one of the highest rates of narrow endemism in the Mediterranean area.

In particular, gorges and the treeless mountain summits are very rich in endemic species. In the mountains of Crete inaccessible cliffs and gorges support a rich endemic chasmophytic flora, which includes *Campanula jacquinii, Dianthus fruticosus, Ebenus cretica* and *Origanum dictamnus*. The high mountain areas of Lefka Ori contain an equally important endemic element, with rare and localized species including *Myosotis solange, Centaurea baldacii, Nepeta sphaciotica* and *Ranunculus radinotrichus* (Turland *et al.*, 1993). In Corsican mountains, although the overall flora is species poor, there are 154 endemic taxa many of which have affinities with alpine-arctic species, dating probably from the late Tertiary (Mouillot *et al.*, 2008).

Mountain-top areas throughout the Mediterranean Basin are refuges for relict conifer tree species (e.g. circum-Mediterranean fir species) as well as for genetically valuable, isolated populations of tree species whose core distribution is located at higher latitudes in temperate regions (e.g. *Pinus sylvestris*). *Abies pinsapo* is a relict species that belongs to the group of circum-Mediterranean fir species, and is endemic to the region on both sides of the Strait of Gibraltar. It forms isolated populations above altitudes of 1000–1200 m on north-facing slopes in coastal mountain ranges of southern Spain (West Baetic range) and northern Morocco (Rif mountains) (Radford *et al.*, 2011).

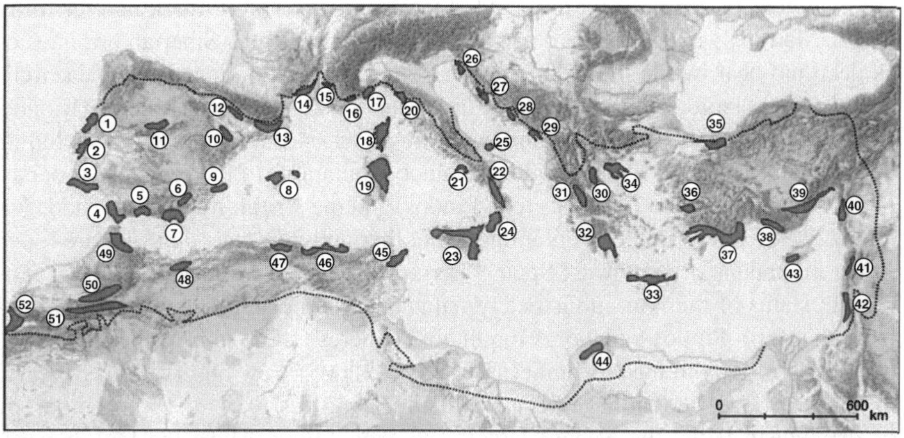

① Beira litoral
② Estramadura
③ Algarve
④ Cadiz/Algeciras region
⑤ Serrania de Ronda
⑥ Sierra Cazorla/Segura
⑦ Sierra Nevada/Gata
⑧ Balearic Islands
⑨ Valencia region
⑩ Ebro Valley
⑪ Sistema central
⑫ S. Pyrenees
⑬ S. E. Pyrenees
⑭ S. Cévennes
⑮ Mont Ventoux
⑯ E. Provence
⑰ Maritime Aips
⑱ Corsica
⑲ Sardinia
⑳ Alpi Apuani
㉑ Campania
㉒ S. Apennines
㉓ Sicilia
㉔ S. Calabria
㉕ Gargano
㉖ N. Istria
㉗ Velebit Mountains
㉘ S. Bosnia/Biokovo
㉙ Montenegro
㉚ Olympe/Katalympos
㉛ C. Greece (Pindos)
㉜ Peloponnese
㉝ Crete
㉞ Chalkidiki peninsula
㉟ Izmit region
㊱ Boz/Aydin dag
㊲ S. W. Anatolia
㊳ C. Taurus
㊴ E. Taurus
㊵ Amanus
㊶ Lebanon range
㊷ Israel/Palestine
㊸ Cyprus
㊹ Cyrenaic (Lybia)
㊺ J. Zaghouan/Cap Bon
㊻ Petite Kabylie/de Collo
㊼ Grande Kabylie
㊽ Tlemcen Mountains
㊾ Rif Mountains
㊿ Middle Atlas
(51) High Atlas
(52) Souss/W. Anti Atlas

Figure 6.5 Species refugia in the Mediterranean. Reproduced from Médail and Diadema (2009), with permission

Médail and Diadema (2009) identified 52 refugia, of which 33 are in the western Mediterranean and 19 in the east. Moreover, half of the refugia correspond with designated hotspots. Médail and Diadema conclude that these refugia are 'significant reservoirs of unique genetic diversity favourable to the evolutionary processes of Mediterranean plant species' and as such should be of high conservation priority. Some 33 of these refugia are in mountain areas, where land of varied habitats between 400 and 800 m asl (and possibly higher) would have provided suitable habitats and allowed treelines to move up and down depending on changes in climatic severity. Examples of mainland mountains are the High and Middle Atlas, the Pyrenees, the Velebit Mountains of Croatia, the Amanus Mountains of southeast Turkey and the Mountains of Lebanon (Figure 6.5).

6.4 Fauna

The mountains of the Mediterranean stand out as areas of high mammal species richness with particularly high concentrations of threatened species found in the mountains of Turkey, the Levant, and northwest Africa. Zoogeographically there are a number of affinities with non-Mediterranean areas (Kryštufek and Griffiths, 1999). Mammals in the northern part of the Basin are of Euro-Siberian origin, such as deer and bear, while in the southern part they are predominantly of Palaearctic origin with a large number of Afro-tropical or Saharo-Sahelian species (Blondel *et al.*, 2010). In the Pyrenees alone there are about 64 mammal species including the brown bear (*Ursus arctos*). Other disjunct populations of the brown bear persist in the mountains of the Hellenids (e.g. Pindos) and the Apennines (Temple and Cuttelod, 2009). A few viable populations of the Iberian lynx (*Lynx pardina*) are found in the mountains of southwest Spain (Sierra Morena and Montes de Toledo). The present status of the lynx in Greece remains uncertain despite reports of sightings from the north Pindos and Voras mountains, whereas there is probably illegal introduction of lynx to the Apennines (Blasi *et al.*, 2005). Few observations of jackal in Greece above 1000 m have been recorded (Giannatos *et al.*, 2005). The deserts of the Judean Hills, the Negev (Israel) and Sinai (Egypt) host the few leopards of the Middle Eastern subspecies (*Panthera pardus jarvisi*), while the Anatolian leopard (*Panthera pardus tulliana*) persists in the western Taurus (Temple and Cuttelod, 2009). Barbary sheep (*Ammotragus lervia*) are dispersed in scattered groups in all chains of the Atlas Mountains, while the globally threatened Cuvier's gazelle (*Gazella cuvieri*) remains in three disjunct areas: the northern Middle Atlas, western High Atlas and Anti-Atlas mountains (Loggers *et al.*, 1992). Other mammals recorded in the High Atlas include the Barbary macaque and hyena (Davis *et al.*, 1994). Mount Lebanon hosts some large carnivores including the golden jackal, wolf, jungle cat and red fox. The forests provide a haven for numerous species, such as the badger, porcupine, squirrel, wild boar, hedgehog, toad, snakes and lizards (Temple and Cuttelod, 2009).

Mediterranean mountains are also home to several endemic species and subspecies of large herbivores, most of which are rare or endangered. The mouflon (*Ovis orientalis*), ancestor of the domestic sheep, is represented by a number of subspecies present in forest areas of Sardinia, Corsica, Cyprus and Turkey, while chamois species are found in central Italy and eastern Anatolia. Several ibex species find refuge on high mountains and rocky outcrops of the Basin including the Nubian ibex (Egypt, Israel, Jordan), the Spanish ibex (Spanish sierras), the Bezoar ibex (Taurus and Anti-Taurus) and the Cretan ibex (Crete) (Davis *et al.*, 1994). In addition, endangered amphibians and reptiles such as the Pyrenean frog (*Rana pyrenaica*) and Aran Rock lizard (*Iberolacerta aranica*) (Cox *et al.*, 2006) are also present.

As with plants, mountains throughout the Mediterranean region have provided refuges for many invertebrates. As a result endemics might account for 15–20% of the total insect fauna in areas like the Atlas, Rif, Pyrenees and Taurus massifs (Blondel *et al.*, 2010). High concentrations of endemic butterfly species are found in the Middle and High Atlas Mountains of Morocco (Thomas and Mallorie, 1985), for example species of the families Pieridae and Hesperiidae, while in some cases evolutionary divergence of butterfly species is favoured by the three-dimensionality of the mountain terrain, as in the case of the genus *Erebia* (Blondel *et al.*, 2010).

Studies on the effects of altitude on butterflies in the Mediterranean have demonstrated that butterfly richness peaks at intermediate levels and quickly decreases at higher altitudes (e.g. Stefanescu *et al.*, 2004). Low richness but high endemism is reported for the ground spiders and other invertebrates in the mountains of Crete (Sfenthourakis and Legakis, 2001; Chatzaki *et al.*, 2005).

In addition, some mountains host a significant number of breeding bird species – in the case of the Pyrenees 120 species – as well as a large number of migratory species (Davis *et al.*, 1994). The high mountain tops such as those of Crete are among the last strongholds of many birds of prey such as the bearded vulture (*Gypaetus barbatus*), the griffon vulture (*Gyps fulvus*), the golden eagle (*Aquila chrysaetus*) and Bonelli's eagle (*Aquila fasciatus*). The forests of the Corsican mountains also host a number of birds such as the endemic Corsican nuthatch (*Sitta whiteheadi*), two species of endangered raptors and rare Palaearctic birds such as the bearded vulture (Mouillot *et al.*, 2008).

6.5 Conservation in Mediterranean mountains

> Mountains are very special places [and] they are the last bastion of wild, untrammelled nature and unfettered evolutionary processes.
>
> Hamilton and McMillan (2004, pp. 1–3)

This statement summarizes and underpins the reasons behind the need for conservation and protected area designation in mountains. The threats to the ecological and functional integrity of mountains that call for conservation action include forestry practices, dams, ski facilities, winter resorts, associated road construction, grazing and land abandonment, with varying degrees of intensity between mountains (see Chapter 8). All these activities do not bode well for mountain biodiversity. Habitat loss and fragmentation are having an impact on native flora and fauna, as in the case of the bear populations in the Spanish Pyrenees or the Greek Pindos mountains.

However, the main challenge for future conservation efforts is climate change and its possible impacts on mountain environments (see Chapter 9). Higher temperatures in mountain regions will lead to an upward shift of biotic zones with possible decrease in the numbers and abundance of endemic species (Mooney *et al.*, 2001; Radford *et al.*, 2011). It is expected that the highest zones (i.e. alti-Mediterranean) will be most affected due to limited possibilities for species upward/altitudinal migration (Médail and Quézel, 2003). Other possible ecosystem responses to climate change, as summarized by Médail and Quézel (2003), include extinction or regression of species populations, migration, northward extension of the thermophilous vegetation, expansion of saharian or sahelian floristic elements in the southern part of the Basin, and evolutionary responses of the vegetation. There is still little empirical evidence of these effects on the Mediterranean mountain flora. For example, recent studies for two Mediterranean mountains, Lefka Ori and the central Apennines, concur on the effect of changes on rare species but indicate that colonization of high altitudes by subalpine species will occur at different paces for the principal exposures (north, east, south, west) (Stanisci *et al.*, 2005; Kazakis *et al.*, 2007). A comparison of current and historical vegetation distribution maps in the Montseny mountains (Catalonia, northeast Spain) demonstrates a progressive altitudinal shift of vegetation zones over a 50-year period (Peñuelas and Boada, 2003). Médail and Verlaque (1997) suggest that the mountain flora of Corsica is in less danger since both endemics and pressures are concentrated at mid-altitudes (800–1700 m). Dolines, pozzines and ponds will be among the most threatened habitats due to changes in hydrological regimes (Ghosn *et al.*, 2010), while screes and rocky cliffs will act as they have in the past as refuges for species (Médail and Quézel, 2003).

Until now conservation practices in mountain areas in general and in the Mediterranean in particular have taken many forms including national parks and natural reserves, natural monuments and protected landscapes. For the Euro-Mediterranean countries two relatively new designations came into force with the implementation of the Natura 2000 network of protected areas from 2000 to 2010 (European Council, 1992), where sites were designated as Special Protection Areas (SPAs) or Special Areas of Conservation (SACs). The percentage of designation in the mountains of the Mediterranean EU countries is shown in Figure 6.6. According to the World Conservation Monitoring Centre (WCMC) database, in many Euro-Mediterranean countries such as Spain, Portugal, Italy and Malta there is a very high overlap between the nationally designated areas (NDAs) and those currently under the Natura

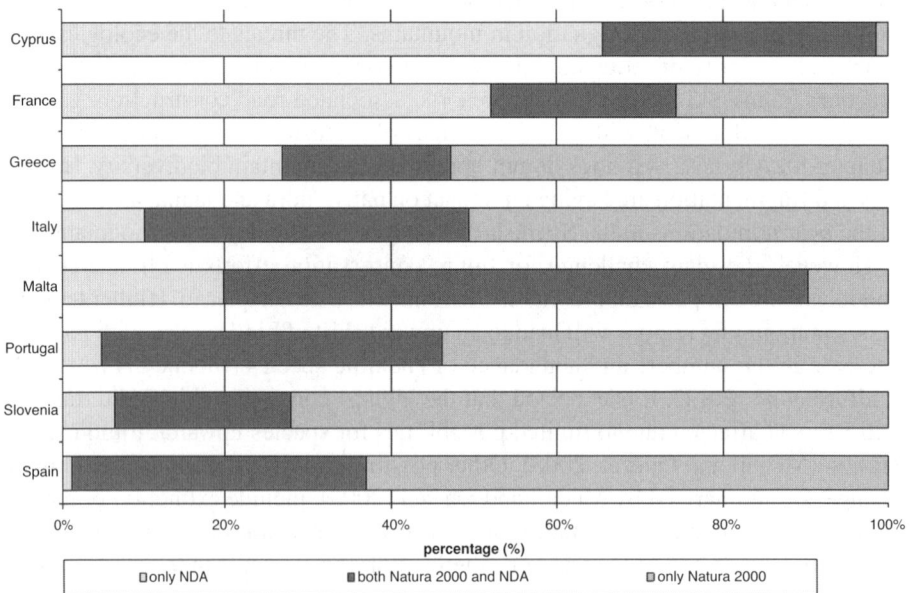

Figure 6.6 Proportion of protected mountain areas that lie within nationally designated areas (NDAs) or the Natura 2000 network, or both, in the Euro-Mediterranean countries (compiled using data from the World Conservation Monitoring Centre database)

2000 network (see Figure 6.6). In other words, protected areas in mountains zones of these countries have a double designation status. Cyprus, France and to a certain extent Greece have still retained a large area of their mountains as national designations only.

Many mountainous areas are part of the UNESCO World Network of Biosphere Reserves. Examples from the Mediterranean include the Sierra Nevada in Spain, and Mount Olympus and Lefka Ori (Samaria Gorge) in Greece. Apart from the 'conventional' in situ designations a relatively recent designation is the Plant Micro-Reserve; this was pioneered in Spain by Laguna et al. (2004) in order to protect species restricted to very small areas. Plant micro-reserves are small areas, typically 1–2 ha, in which there is a significant presence of rare, threatened or endemic plant species. The management of small areas is relatively simple and can easily be adapted when necessary. Currently micro-reserves have been established in the Lefka Ori massif of Crete and the Troodos Mountain in Cyprus. The majority of the designations mentioned so far are species/habitat centred. However, for mountain areas the IUCN Category IV (Protected Landscapes) is often advocated (IUCN 1994) since it has the flexibility to offer protection to the whole area while at the same time providing specific protection for specially defined purposes (e.g. habitats/species) as seen necessary.

6.6 Conclusion

In the Mediterranean Basin with its long history of human activity, mountains are considered to be some of the last remaining wilderness areas with high landscape and biodiversity value. Despite the harsh environment, mountain ecosystems provide a wide range of ecosystem services (cf. MEA, 2005) including water, timber, habitat provision, recreation and carbon sequestration, and for these reasons they are important internationally.

Geology, tectonic activity, isolation and limited human activity explain the current biogeographical patterns occurring in Mediterranean mountains. These patterns conform to theory, with mountains displaying low species richness but high endemism along altitudinal gradients, although there are some exceptions. Mountains have played a refuge role in geological history and to a certain extent they retain this role today. Despite adversities there are still pristine areas in many Mediterranean mountains that sustain a diversity of plant species but also populations of large mammals and birds of prey. However, this role is impeded by ongoing human activities and most importantly climate change. For this reason conservation should not focus simply on designation and protection but on the design of a flexible reserve system along with conventional ex situ conservation measures. Recent conservation efforts worldwide focus on the permeability of the intervening landscape matrix, dispersal corridors and habitat networks (Jongman and Pungetti, 2004; Watts and Handley, 2010), and these should also be applied in the case of mountains. Moreover, and despite the fact that past designations in mountain areas have focused on the protection of physical, biological and aesthetic ecosystem qualities, the preservation of community livelihood and culture in these areas is now considered central to conservation efforts (Hamilton and McMillan, 2004) and in agreement with the ecosystem-based approach (MEA, 2005).

References

References marked as bold are key references.

Benedetto, G. and Giordano, A. (2008) Sicily. In: Vogiatzakis, I.N., Pungetti, G. and Mannion, A. (eds) *Mediterranean Island Landscapes: Natural and Cultural Approaches*. Landscape Series Vol. 9. Springer Publishing, pp. 117–142.

Blasi, C., Boitani, L., La Posta, S., Manes, F. and Marchetti, M. (2005) *Stato della biodiversita in Italia: contributo alla strategia nazionale per la biodiversita*. Palombi Editori.

Blondel, J., Aronson, J., Bodiou, J-Y. and Boeuf, G. (2010) *The Mediterranean Region: Biological Diversity through Time and Space*. Oxford University Press.

Chatzaki, M., Lymberakis, P. and Mylonas, M. (2005) The distribution of ground spiders (Araneae, Gnaphosidae) along the altitudinal gradient of Crete, Greece: species richness, activity and altitudinal range. *Journal of Biogeography* 32:813–831.

Cox, N., Janson, J. and Stuart, S. (eds) (2006) *The Status and Distribution of Reptiles and Amphibians in the Mediterranean Basin*. IUCN.

Davis, S.D., Heywood, V.H. and Hamilton, A.C. (eds) (1994) *Centres of Plant Diversity*. Cambridge: WWF/IUCN.

Davis, S.D., Heywood, V.H. and Hamilton, A.C. (eds) (1999) *Centres of Plant Diversity, Volume 1: Europe, Africa, South West Asia and the Middle East: A Guide and Strategy for their Conservation*. IUCN.

Dominguez, F., Galicia, D., Moreno-Rivero, L., Moreno Saiz, J.C. and Sainz de Ollero, H. (1996) Threatened plants in Peninsular and Balearic Spain: a report based on the EU Habitats Directive. *Biological Conservation* 76:123–133.

Egli, B. (1989) Ecology of dolines in the mountains of Crete. Bielefelder ökologische Beiträge 4:59–63.

Egli, B. (1991) The special flora, ecological and edaphic conditions of dolines in the mountains of Crete. *Botanika Chronika* 10:325–335.

European Council (1992) Council Directive 92/43/EEC of 21 May 1992 on the conservation of natural habitats and of wild fauna and flora. OJ L 206, 22.7.1992, pp. 7–50.

Favarger, C. (1972) Endemism in the montane floras of Europe. In: Valentine, V.H. (ed.), *Taxonomy, Phytogeography and Evolution*. London: Academic Press, pp. 191–204.

Ghosn, D., Vogiatzakis, I.N., Kazakis, G. *et al.* (2010) Ecological changes in the highest temporary pond of western Crete (Greece): past, present and future. *Hydrobiologia* 648:3–18.

Giannatos, G., Marinos, Y., Maragou, P. and Catsadorakis, G. (2005) The status of the Golden Jackal (*Canis aureus* L.) in Greece. *Belgian Journal of Zoology* 135:145–149

Gómez, D., Sesé, J.A. and Villar, L. (2003) The Vegetation of the Alpine Zone in the Pyrenees. In: Nagy, L., Grabherr, G., Korner, C. and Thompson, D.B.A. (eds), *Alpine Biodiversity in Europe*. Ecological Studies Vol. 167. Springer.

Greuter, W. (1972) The relict element of the flora of Crete and its evolutionary significance. In: Valentine, V.H. (ed.), *Taxonomy, Phytogeography and Evolution*. London: Academic Press, pp. 161–177.

Hamilton, L.S. and McMillan, L. (eds) (2004) *Guidelines for Planning and Managing Mountain Protected Areas*. Gland and Cambridge: IUCN.

IUCN (1994a) *Guidelines for Protected Area Management Categories*. CNPPA with the assistance of WCMC. IUCN, Gland, Switzerland and Cambridge, UK.

Jongman, R.H.G. and Pungetti, G. (2004) *Ecological Networks and Greenways: Concept, Design, Implementation*. Cambridge Studies in Landscape Ecology. Cambridge University Press.

Kaya, Z. and Raynal, D.J. (2001) Biodiversity and conservation of Turkish forests. *Biological Conservation* 97:131–141.

Kazakis, G., Ghosn, D., Vogiatzakis, I.N. and Papanastasis, V.P. (2007) Vascular plant diversity and climate change in the alpine zone of the Lefka Ori, Crete. *Biodiversity and Conservation* 16:1603–1615.

Körner, C. and Spehn E.M. (eds) (2002) *Mountain Biodiversity: A Global Assessment*. Parthenon Publishing.

Kruckenberg, A.R. and Rabinowitz, D. (1985) Biological aspects of endemism in higher plants. *Annual Review of Ecology and Systematics* 16:447–479.

Kryštufek, B. and Griffiths, H.I. (1999) Mediterranean v. continental small mammal communities and the environmental degradation of the Dinaric Alps. *Journal of Biogeography* 26:167–177.

Laguna, E., Deltoro, V.I., Pèrez-Botella, J. *et al.* (2004) The role of small reserves in plant conservation in a region of high diversity in eastern Spain. *Biological Conservation* 119:421–426.

Loggers, C., Thévenot, M. and Aulagnier S. (1992) Status and distribution of Moroccan wild ungulates. *Biological Conservation* 59:9–18.

Major, J. (1988) Endemism: a botanical perspective. In: Myers, A.A. and Giller, P.S. (eds), *Analytical Biogeography*. London: Chapman & Hall, pp. 117–146.

MEA (Millennium Ecosystem Assessment) (2005) *Ecosystems and Human Well Being*. Island Press.

Médail, F. and Diadema, K. (2009) Glacial refugia influence plant diversity patterns in the Mediterranean Basin. *Journal of Biogeography* 36:1333–1345.

Médail, F. and Quézel, P. (1997) Hot-spots analysis for conservation of plant biodiversity in the Mediterranean Basin. *Annals of the Missouri Botanical Gardens* 84:112–127.

Médail, F. and Quézel, P. (2003) Conséquences écologiques possibles de changements climatiques sur la flore et la végétation du basin méditerranéen. *Bocconea* 16:397–422.

Médail, F. and Verlaque, R. (1997) Ecological characteristics and rarity of endemic plants from southeast France and Corsica: implications for biodiversity conservation. *Biological Conservation* 80:269–271.

Mooney, H.A., Kalin Arroyo, M.T., Bond, W.J. *et al.* (2001) Mediterranean-climate ecosystems. In: Chapin, F.S., III, Sala, O.E. and Huber-Sannwald, E. (eds) *Global diversity in a changing environment. Scenarios for the twentyfirst century*. Ecological Studies 152. New York: Springer, pp. 157–199.

Moore, H.M., Fox, H.R., Harrouni, M.C. and El Alami, A. (1998) Environmental challenges in the Rif mountains, northern Morocco. *Environmental Conservation* 25:354–365.

Mota, J.F., Perez-Garcia, F.J., Jimenez, M.L., Amate, J.J. and Penas, J. (2002) Phytogeographical relationships among high mountain areas in the Baetic ranges (South Spain). *Global Ecology and Biogeography* 11:497–504.

Mouillot, F., Paradis, G., Andrei-Ruiz, M-C. and Quilichini, A. (2008) Sicily. In: Vogiatzakis, I.N., Pungetti, G. and Mannion, A. (eds) *Mediterranean Island Landscapes: Natural and Cultural Approaches*. Landscape Series Vol. 9. Springer Publishing, pp. 220–244.

Myers, N., Mittermeier, R.A., Mittermeier, C.G., da Fonseca, G.A.B. and Kent. J. (2000) Biodiversity hotspots for conservation priorities. *Nature* 403:853–858.

Nagy, L. and Grabherr, G. (2009) *The Biology of Alpine Habitats.* **Oxford University Press.**

Nagy, L., Grabherr, G., Korner, C. and Thompson, D.B.A. (eds) (2003) *Alpine Biodiversity in Europe.* Ecological Studies Vol. 167. Springer.

Parolly, G. (2004) The high mountain vegetation of Turkey – a state of the art report, including a first annotated conspectus of the major syntaxa. *Turkish Journal of Botany* 28:39–63.

Pedrotti, F. and Gafta, D. (2003) The high mountain flora and vegetation of the Apennines and the Italian Alps. In: Nagy, L., Grabherr, G., Korner, C. and Thompson D.B.A. (eds), *Alpine Biodiversity in Europe.* Ecological Studies Vol. 167. Springer, pp. 73–84.

Peñuelas, J. and Boada, M. (2003) A global change-induced biome shift in the Montseny mountains (NE Spain). *Global Change Biology* 9:131–140.

Pignatti, S. (1978) Evolutionary trends in Mediterannean flora and vegetation. *Vegetation* 37:175–185.

Quézel, P. (1981a) Floristic composition and phytosociological structure of sclerophyllous mattoral around the Mediterranean. In: Di Castri, F., Goodall, D.W. and Sprecht, R.L. (eds), *Mediterranean-Type Shrublands.* Amsterdam: Elsevier, pp. 107–121.

Quézel, P. (1981b) The study of groupings in the countries surrounding the Mediterranean: some methodological aspects. In: Di Castri, F., Goodall, D.W. and Sprecht, R.L. (eds), *Mediterranean-Type Shrublands.* Amsterdam: Elsevier, pp. 87–93.

Quezel, P. (1981c) Les hautes montagnes du maghreb et du proche-orient: essai de mise en parallèle des charactères phytogéographiques. *Anales de Jardin Botanico de Madrid* 37:353–372.

Quézel, P. (1985) Definition of the Mediterranean region and the origins of the flora. In: Gomez-Campo, C. (ed.), *Plant Conservation in the Mediterranean Area.* Dordrecht: Dr Junk, pp. 9–24.

Quézel, P. (2004) Large-scale post-glacial distribution of vegetation structures in the Mediterranean region. In: Mazzoleni, S., di Pascuale, G., Mulligan, M., di Martino, P. and Rego, F. (eds), *Recent Dynamics of the Mediterranean Vegetation and Landscape.* **Chichester: John Wiley & Sons, Ltd, pp. 3–12.**

Quézel, P. and Médail, F. (2003) *Ecologie et Biogéographie des Forêts du Bassin Méditerranéen.* **Paris: Elsevier, Collection Environnement.**

Radford, E.A., Catullo G. and de Montmollin B. (2011) *Important Plant Areas of the South and East Mediterranean Region: Priority Sites for Conservation.* IUCN, Plant Life, WWF.

Sfenthourakis, S. and Legakis, A. (2001) Hotspots of endemic terrestrial invertebrates in southern Greece. *Biodiversity and Conservation* 10:1387–1417.

Snogerup, S. (1971) Evolutionary and plant geographical aspects of chasmophytic communities. In: Davis, P.H., Harper, P.C. and Hedge, I.G. (eds), *Plant Life of South West Asia.* Edinburgh: The Botanical Society of Edinburgh, pp. 157–169.

Stanisci, A., Pelino, G. and Blasi, C. (2005) Vascular plant diversity and climate change in the alpine belt of the central Apennines (Italy). *Biodiversity and Conservation* 14:1301–1318.

Stefanescu, C., Herrando, S. and Páramo, F. (2004) Butterfly species richness in the north-west Mediterranean Basin: the role of natural and human-induced factors *Journal of Biogeography* 31:905–915.

Strid, A. (ed.) (1986) *Mountain Flora of Greece 1.* Cambridge: Cambridge University Press.

Strid, A. (1993) Phytogeographical aspects of the Greek mountain flora. *Fragmenta Floristica et Geobotanica Supplement* 2(2):411–433.

Strid, A. (1995) The Greek mountain flora, with special reference to the Central European element. *Bocconea* 5:99–112.

Strid, A. and Papanikolaou, K. (1985) The Greek mountains. In: Gomez-Campo, C. (ed.), *Plant Conservation in the Mediterranean Area*. Dordrecht: Dr Junk, pp. 89–111.

Strid, A. and Tan, K. (eds) (1991) *Mountain Flora of Greece 2*. Edinburgh: Edinburgh University Press.

Temple, H.J. and Cuttelod, A. (compilers) (2009) *The Status and Distribution of Mediterranean Mammals*. Gland/Cambridge: IUCN.

Terrab, A., Hampe, A., Lepais, Talavera, S., Vela, E. and Stuessy, T.F. (2008) Phylogeography of North African Atlas cedar (*Cedrus atlantica*, Pinaceae): Combined molecular and fossil data reveal a complex Quaternary history. *American Journal of Botany* 95:1262–1269.

Thomas, C.D. and Mallorie, H.C. (1985) Rarity, species richness and conservation: butterflies of the Atlas Mountains in Morocco. *Biological Conservation* 33:95–117.

Tivy, J. (1993) *Biogeography: A Study of Plants in the Ecosphere*, 3rd edn. Harlow: Longman.

Tsintides, C.T., Hadjikyriakou, N.G. and Christodoulou, C.S. (2002) *Trees and Shrubs in Cyprus*. Nicosia: Leventis Foundation and Cyprus Forest Association.

Turland, N.J., Chilton, L. and Press, J.R. (1993) *Flora of the Cretan area: Annotated Checklist and Atlas*. London: The Natural History Museum, HMSO.

Tvrtkoviç N. and Veen P. (eds) (2006) *The Dinaric Alps: Rare Habitats and Species. A Nature Conservation Project in Croatia*. Zagreb: Hrvatski prirodoslovni muzej (CNHM) and Royal Dutch Society for Nature Conservation (KNNV).

Vogiatzakis, I.N. (2000) Predicting the distribution of plant communities in Lefka Ori, Crete, using GIS. Unpublished PhD thesis, University of Reading.

Vogiatzakis, I.N. and Griffiths, G.H. (2006) A GIS-based empirical model for vegetation prediction in Lefka Ori, Crete. *Plant Ecology* 184:311–323.

Vogiatzakis, I.N., Griffiths, G.H. and Mannion, A.M. (2003) Environmental factors and vegetation composition of the Lefka Ori massif, Crete, S. Aegean. *Global Ecology and Biogeography* 12:131–146.

Vogiatzakis, I.N., Mannion A.M. and Griffiths, G.H. (2006) Mediterranean Ecosystems: problems and tools for conservation. *Progress in Physical Geography* 30:175–200.

Watts, K. and Handley, P. (2010) Developing a functional connectivity indicator to detect change in fragmented landscapes. *Ecological Indicators* 10:552–557.

White, F. (1983) *The Vegetation of Africa*. Paris: UNESCO.

Zaffran, J. (1990) *Contributions à la Flore et à Vegétation de la Crète*. Aix en Provence: Universitè de Provence.

7

Cultural geographies

Veronica della Dora and Theano S. Terkenli

7.1 Introduction

> Geology explains the overabundance of mountains across the solid space of the Mediterranean. Recent, high, and craggy mountains; mountains which, like a stony skeleton, pierce the skin of the Mediterranean territory: the Alps, the Apennines, the Balkans, Taurus, Lebanon, Atlas, Spain's mountain ranges, the Pyrenees – what a parade! Steep snow-capped peaks, towering above the sea and above planes where roses and orange trees blossom; abrupt promontories which often terminate in the water: these are landscapes we find almost interchangeably from one shore of the Mediterranean to the other.
>
> Braudel (2000, pp. 14–15)

The Mediterranean, it has been argued, 'is not so much the sea between the lands, as the name asserts, but the sea among the mountains' (McNeill, 1992, p. 12). Sailing around the Mediterranean Basin (except for part of the North African coast), one is practically almost never out of sight of a range or peak. Fernand Braudel famously narrated Mediterranean mountains in terms of continuity: from above, as an ensemble of ranges that bound the watery continent for most of its perimeter; from the ground, or sea level, as 'interchangeable' landforms framed by an iconic landscape of water, roses and orange trees. However, it is not until the age of steam that sailors would have ventured directly from one side of the Basin to the other. The Mediterranean of antiquity was not a single homogeneous watery continent diametrically criss-crossed by merchant routes, but rather a mosaic of ecological 'microregions' connected through *cabotage*, or coastwise navigation (Horden and Purcell, 2001).

The open sea's flat horizon was feared. Coastal landmarks, especially promontories and lofty mountain peaks, by contrast, were deemed reassuring reference points for sailors. Indeed, outbound travellers would often be able to discern the faint silhouette of distant lands looming on their visual horizon even before leaving their port. The mountains of Epirus were visible to sailors from the Italian shore across the straits of Otranto. The snow-capped peaks of Sierra Nevada on the southern

Mediterranean Mountain Environments, First Edition. Edited by Ioannis N. Vogiatzakis.
© 2012 John Wiley & Sons, Ltd. Published 2012 by John Wiley & Sons, Ltd.

coast of Spain rose for over 3400 m asl commanding a view of the Moroccan range opposite, beckoning 'the first Phoenician trireme which felt its way westward along the African littoral', and partly accounting 'for the constant intercourse between these neighbouring coasts from the earliest times' (Semple, 1932, p. 589). In clear atmospheric conditions, Phoenicians pruning their vineyards on the terraced slopes of Lebanon would be able to gaze at the 1952 m-high peak of Olympus in Cyprus, from whose heights one would in turn be able to see the mighty Taurus range in Asia Minor.

Remarkable Aegean peaks served both as beacons and as meteorological stations. 'A turban cloud on Mt. Oros on the island of Aegina was a sign of rain.... A mantle of clouds about Athos presaged a storm; a girdle of clouds half way up its slope indicated a southerly wind and eventual rain' and so on (Semple, 1932, pp. 521–22) (Figure 7.1a). Thanks to their visibility and physical specificities, each

Figure 7.1   (a) View of Mount Athos (photograph by Monk Apollò Docheiarite). (b) Fischer von Erlach's representation of the Mount Athos colossus as envisaged by Dinocrates, in *Entwurf einer historischen Architectur*, Leipzig, 1712 (courtesy of the Gennadius Library, American School of Classical Studies at Athens)

of these peaks would have been at once familiar and unique to sailors – not only as practical landmarks for navigation, but also as cultural landmarks often associated with specific divinities. In Byzantine times, many of the eastern Mediterranean peaks underwent resignification and became privileged abodes of monks and hermits. Under the Ottomans, some of them were reinscribed with mosques and stories of brigands, adding further layers to the pre-existing ones. Today most of these summits are unique biogeographical observatories, as the previous chapter showed. They are also privileged destinations for skiers, hikers and other seekers of nature who, perhaps without their realizing it, are after a spiritual experience somehow akin to that of early pilgrims venturing to the same summits.

Stories of cultural resignification and overlayering repeat themselves on all shores of the Mediterranean, leaving both visible and imaginative marks on the landscapes of its high places: pagan temples, Christian chapels, mosques, winter resorts, variously rooted placenames and, not least, a plethora of stories. This chapter provides a brief introduction to some of these overlayerings. While mountains possess great metaphorical potential, they remain insistently material objects. Cultural geographers are interested in materiality as it impacts geographical imagination and embodied practices (see, e.g., Cosgrove and della Dora, 2009; Debarbieux, 1998, 2004). 'Moralized', 'politicized', even 'sanctified', to be sure, but also 'lived from within', Mediterranean mountains offer a unique opportunity for examining substantive examples of the active intertwining of natural objects with human subjectivity and cultural specificities. Accordingly, the chapter is articulated through four different 'narrative layers'. Following a broadly chronological order, these engage respectively with myth, religion, tradition and commodification. The chapter opens with mountain myths and pagan sanctuaries in the ancient Mediterranean world. The following section explores mountain theophanies in the great monotheistic traditions of the Mediterranean: Judaeo-Christianity and Islam. It shows how Biblical mountain imageries migrated from Sinai and Palestine to other Mediterranean shores, creating new networked mountain geographies. The following two sections explore more contemporary narratives of Mediterranean high places, as they evolve out of age-old ways of life, adjusted to difficult geographies, but drawing out of their material particularities, natural resources and deep symbolism to create new, viable forms of livelihood. While by no means comprehensive, together the four sections aim at reflecting the complex overlayering and transformation of the cultural significance of Mediterranean mountains through the centuries and at opening up questions about their future.

7.2 Mythical mountains

Mountains have an extraordinary power to evoke the sacred. The ethereal rise of a ridge in the mist, the glint of moonlight on an icy face, a flare of gold from a distant peak – such glimpses of transcendent beauty can reveal our world as a place of unimaginable mystery and splendour.

Bernbaum (1997, p. xiii)

While Mediterranean peaks do not reach Andean and Himalayan heights, the visual contrast produced by their verticality and the surrounding landscape, accentuated by the brightness of the southern light, has made them objects of awe since the dawn of civilization. Faith and lore have further elevated these summits, while visual interconnectedness made them nodes of cultural networks, providing ancient cults and myths with both geographical and narrative continuity. Holy mountains are attested from the Bronze Age in the religious traditions of the Levant and were central features in Minoan Crete (Horden and Purcell, 2001, p. 413). The Minoan civilization flourished on the island between 3000 and 1500 BC. Its religion centred on the cult of the 'mountain mother', a female goddess associated with the fertility of the land and worshipped on high places.

Neither too high to be forbidding nor too low to pass unobserved, Cretan peaks offered relatively easy points of access to the divine. More than 50 mountain-top shrines are present on the island and could be reached within a 3-hour walk from the main settlement. Almost all of them are located in altitudinal regions associated with the summer transhumance of sheep and goats, perhaps 'to relieve the fears and cares of the shepherds and breeders' (Rutkowski, 1986, p. 185). More significantly, the sanctuaries are set in sight of each other, and in sight of the villages in the valley below (Peatfield, 1983; Rackham and Moody, 1996). Large sacrificial bonfires were lit as part of ceremonies, providing spiritual comfort to those villagers in the valley who would lift their gaze to heaven. During festival nights a network of sacred beacons would unite various regions and allow the faithful to perform a 'visual' pilgrimage through the peaks (Peatfield, 1983, p. 277).

At the close of the Middle Bronze Age, Minoan peak sanctuaries were abandoned for caves. The catastrophic eruption of Thera (Santorini) in the second millennium BC one of the largest volcanic events on Earth ever recorded in history, which devastated the island as well as coastal regions of Crete, caused Minoans to turn from the powerless gods of the skies, from where the poisonous ashes came, to chthonic divinities that might put an end to earthquakes. At the same time, successive waves of invaders from the Eurasian interior had gradually started to penetrate Greece, introducing a religion dominated by masculine deities ruled by a god of thunder and lightning. Out of the encounter between the culture of these tribes and that of the Minoans, the classical civilization of Ancient Greece was born – and with it new sacred mountains and mountain myths (Bernbaum, 1997, p. 106; Rutkowski, 1986, p. 201).

In ancient Greece, each typology of landscape with pronounced physical properties became a manifestation of a particular divinity (Norberg-Schultz, 1979, pp. 29–31). Separated from mankind and yet still in the world of humans, Greek gods found mountains congenial dwellings between earth and heaven. As with Minoan peak shrines, many Hellenic sanctuaries are famous for their spectacular cliff-side or mountain-top settings: 'there seems to have been an idea that the gods needed to live where they could gaze down upon the world' (Williams, 1989, p. 79). As the highest summit, snow-capped Thessalian Olympus (2917 m asl) hosted the throne of Zeus and the 12 gods. Further south, Mount Parnassus (2457 m asl) was deemed to be the dwelling of the Muses. On its slopes was Delphi, the sanctuary of Apollo.

While sanctuaries dedicated to Zeus, the god of rain, thunder and lightning, were located on mountain tops, shrines dedicated to Apollo did not need to stand above the clouds. They were nevertheless usually built in dramatic elevated spots. The sharp contrast between the bright geometrical forms of the shrine and the rough masses of the earth exalted the nature of the young god of light and reason 'dramatizing at once the terrible scale of nature and the opposing patterns which are the result of disciplined human action in the world' (Scully, 1979, p. 100).

As with Minoan peaks, visibility remained a crucial aspect of the sacred geographies of ancient Greece. These geographies, in turn, often remained closely intertwined with coastal navigation and its 'practical geographies'. In ancient Greek the word *oros* means both 'mountain' and 'landmark', a testimony to the enduring importance of mountains and promontories as landmarks for sailors (Figure 7.1a). Templed promontories are to be found not only in the Aegean, but also on all the shores of the Mediterranean and beyond: from the tip of Sinai, 'a notorious spot for conflicting winds and currents dreaded by Greek and Roman seamen', to southern Portugal and the Crimean Bosporus (Semple, 1932, p. 614). These holy seamarks signalled places of severe storms, as the high relief on which the sanctuaries were located often converted the straits and bays below into sea canyons through which winds seasonally blew with restless violence. Templed promontories marked the fine line between known land and boundless sea (Mavian, 1992, p. 41). As such, they were also the first familiar features the returning seaman would have glimpsed on return to his homeland.

Lofty mountain peaks were also employed as beacons to convey long-distance messages. 'The Assyrians used beacons at fixed distances of two hours' journey, and, since lighting a beacon of itself can carry no detailed message, a fast courier was dispatched with the news at the same time' (Pattenden, 1983, p. 269). Likewise, according to Herodotus (*History*, 9: 3), during their invasion of Greece in 480 BC the Persians set up an efficient system, extending from the coast of Asia Minor across the Aegean islands to Attica. The employment of fire signals by the Greeks became common by the time of the Peloponnesian War (431–404 BC). The Macedonians imitated the Persians, and were in turn imitated by the Romans (Pattenden, 1983, pp. 269–70). The most notorious account of mountain beacon-signalling is given by Aeschylus in his description of the capture of Troy by the Greeks. The news of victory was conveyed almost instantly through a chain of mountain beacons linking Troy to Argos, which corresponded with sea marks that had been long familiar to Greek sailors (*Agamemnon*, 281–316; Semple, 1932, pp. 586–88).

In the Mediterranean of classical antiquity, however, mountains were not only locations of sanctuaries and beacons. Visually striking peaks were often repositories of titanic myths used to explain natural phenomena. Mount Etna, the highest volcano in Europe, was deemed to be the residence of Hephaestus. Here the deity of fire and forge would place his heavy anvil on Typhon, a monster Zeus had imprisoned in the fiery mountain. 'At every eruption the earth groaned and shook, owing to the movements of the wounded giant' (Semple, 1932, p. 52). Far across the Mediterranean, Atlas was sentenced by the President of the Immortals to stand in endless anguish to support the weight of the sky on his shoulders and, according to the Romans, turned into the Atlas range in North Africa, rising to 4167 m asl above the Strait of Gibraltar, the westernmost limit of the inhabited world of antiquity. Mount Athos, the most remarkable summit in the north Aegean and home to a famous Apollo sanctuary, is ascribed similar origins. According to ancient mythology, Athos was a Thracian titan who hurled that whole rocky mass at Poseidon in a clash between giants and gods (Kadas, 1998, p. 9). Due to its dramatic morphology, Athos endured as a privileged site of titanic visions as late as the first century BC. According to the Roman architectural writer Vitruvius the peak captured the imagination of Dinocrates, Alexander the Great's architect, who proposed to carve it into a colossal human figure – by implication that of his mighty patron (*De Architectura*, 2: 1–2) (Figure 7.1b). Widely debated in Renaissance Europe, the Dinocratic myth inspired (similarly unrealized) bold projects across the Mediterranean: from young Michelangelo's plan to carve a colossus into the marble cliffs of Carrara to Greek-American sculptor Papadopoulos's contemporary plan to convert Mount Kerdyllion, a craggy hill in the Macedonian countryside, into a huge head of Alexander the Great – the ultimate irony for the king who declined Dinocrates' vision, and the ultimate testimony to the enduring power of rock on human imagination (Schama, 1995, p. 404; della Dora, 2005). The physical distinctiveness of Mediterranean mountains as 'landmarks', however, did not only capture the imagination of ancient visionaries; as the following section will show, it was destined to become central to the three great monotheistic religions that shaped the spiritual history of the Basin.

7.3 Theophanic mountains

> The mountain occurs among the images that express the connection between heaven and earth; . . . hence [it] marks the highest point in the world.
>
> Eliade (1959, pp. 37–38)

Judaic, Christian and Muslim traditions are signposted by mountain revelations – most of them occurring on peaks around the Mediterranean. Old Testament prophets all encountered God in high places: Ararat, Moriah, Horeb, Carmel, Sinai and Zion. Peter, James and Jacob were blessed with the vision of the transfigured Christ on the top of Tabor. Muhammad received his first revelation from the archangel Gabriel in a cave on Jabal al-Nur ('the Mountain of Light'), near Mecca, and ascended to heaven from Mount Zion in Jerusalem (where the Dome of the Rock stands today). In earliest Hebrew cosmology mountain tops were the closest spots to the Upper Chambers above the firmament, where God was believed to reside and come down to meet with the faithful (Psalm 104: 13). As such, they endured in the three monotheistic traditions as the closest earthly places to God's abode and thus as the natural settings for theophanies.

The history of the three religious traditions can be read as a cumulative process. For Christians, Old Testament prophecies were fulfilled by Christ, the Messiah and Son of God; Judaism is therefore regarded as the 'first act' of the history of human salvation. Muslims in turn envisage their religion as the completed and universal version of the same primordial, monotheistic faith revealed at different times through different prophets, starting with Abraham and Moses, continuing with Jesus, and culminating with Muhammad. For this reason, Old Testament theophanic sites are considered sacred in all the three traditions and have often been contested places. Among the plethora of Old Testament high places (see Semple, 1932, p. 519) Sinai and Zion stand out across the three traditions. Separated by nearly 400 km, these mountains embody opposite and yet complementary poles of Judaic tradition. Craggy, barren, awe-inspiring, Sinai is the peak of the covenant and the law. On the summit of this 2288 m-high massif located in the southern part of the desert peninsula after which it is named (one of the most arid regions of the world), Moses received the Commandments 'in a dark cloud of thick smoke' (Exodus 19: 9; 1 Kings 19: 8–13). Zion, on the other hand, was the beautiful site of the Temple and the priest, the capital city and mountain of the kings (Bernbaum, 1997, p. 102). The opposed physical geographies of the two sites can be interpreted metaphorically:

> In the history of Judaism, Sinai, the rugged peak of the wilderness, gives way to Zion, the cultivated mountain of civilization. . . . The voice that sounded in the open space of the desert now echoes in the narrow streets of [Jerusalem]. Zion incorporates and fulfils in human society the meaning of the lonely encounter on Sinai. Sinai is the mountain of the beginning, Zion the mountain of the end.
>
> Bernbaum (1997, p. 102)

The Sinai-Zion dichotomy can also be read within the history of Christianity and in the schism that in 1054 AD split the Christian Mediterranean between the Latin and Eastern Churches: the former making Mount Zion a model for the Vatican in Rome with the pope, whose authority the Eastern branch rejected, as the 'high priest of Christianity'; the latter identifying with Sinai a prefiguration of Tabor and the mystical experience of *theosis*, or union with God, to which every Orthodox Christian is called (Bernbaum, 1997, p. 103).

Here Tabor emerges as a third pole. Rising to an altitude of no more of 575 m asl, the lofty hill pops up almost unexpectedly from the flat landscape of lower Galilee (Figure 7.2a). As with other Mediterranean holy peaks of antiquity, the visual contrast is enough to justify its biblical epithet of 'high mountain' (Matthew 17: 1). Its modest altitude and gentle shape are elevated and dramatized by Byzantine iconography (Figure 7.2b). On icons of the Transfiguration, Tabor is usually portrayed as a craggy peak topped by the transfigured Christ. The mountain is in turn transfigured in the two Old Testament peaks (Horeb topped by Elijah on the left and Sinai

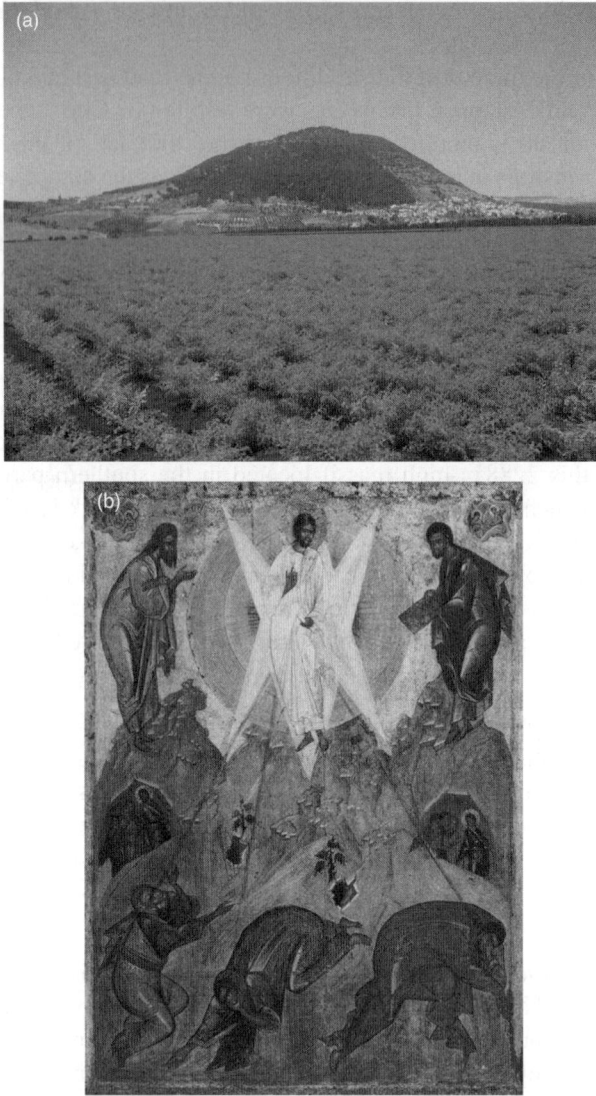

Figure 7.2   (a) View of Mount Tabor (source: Wikimedia). (b) Theophanes the Greek, Transfiguration, fourteenth century (source: Wikimedia) (© Tretyakov Gallery, Moscow). (*A full colour version of this figure appears in the colour plate section*)

topped by Moses on the right), indicating the fulfilment of both the Law and the Prophets (Matthew 5: 17; Nes, 2007).

Today the peak of Sinai is topped by the Greek-Orthodox chapel of Moses flanked by a minaret, and that of Zion by the Dome of the Rock, or Mosque of Omar (built on the site of the Temple of Solomon), ranking as the third most important pilgrimage destination in Islam after Mecca and Medina (Bernbaum, 1997,

p. 100). On Tabor stands the Franciscan church of the Transfiguration, built on the top of the remains of an early Byzantine church and a Crusader church, whereas chapels of the Transfiguration situated on high peaks (often on the sites of pagan temples) punctuate the cultural landscape of Greece and the eastern Mediterranean.

Christian holy mountains are not limited to the theophanic sites mentioned in the Scriptures. Between the ninth and the eleventh centuries, remarkable biblical peaks set a pattern for a number of brand new holy mountains in the eastern Mediterranean. With Arab invasions pushing anchorites out of the Egyptian and Palestinian deserts, various non-biblical holy peaks started to emerge in the Byzantine Empire: Auxentius, Latros and Kyminàs in Bithynia; Olympus in Mysia; Galesion near Ephesus; Ida off the coast of Lesbos, Ganos and Paroria in Thrace, Athos in the Khalkidiki, the Wondrous Mountain in Syria, and so on. Initially, these mountains were attributed an aura of holiness because of the presence of charismatic holy men, and later because of the establishment of organized monastic communities (Talbot, 2001, p. 264). Although, unlike Sinai or Tabor, they were not linked with biblical theophanies, these mountains (some of which had been sites of ancient pagan myths) often ended up surpassing their scriptural counterparts in fame. For example, in the tenth century Mount Athos became *the* Holy Mountain of Orthodoxy, a title that it continues to retain to the present, along with its 20 Byzantine monasteries (Talbot, 2001, p. 269).

Christian ascetics in search of spiritual quietness were attracted to these mountains because of their geographical separation from society, their detachment from normal conditions of life, and the numinous quality of their craggy landscapes. As with the sacred peaks of ancient Greece and biblical mountains, Byzantine holy mountains did not stand in total isolation from each other, nor from society. They constituted nodes of extensive sacred networks and operated as focal centres of spiritual resource and pilgrimage, as well as strongholds of Orthodox faith. Separated, yet still accessible from the main urban centres, some of these holy mounts had a significant influence on the world at their feet: peasants, generals and emperors alike sought the prayers and advice of their residents, whereas clergy or politically involved laymen persecuted by civil authorities often found a safe refuge in their monasteries (Greenfield, 2000).

When a hermit or a community moved to a deserted mountain for the first time, besides the harsh living conditions dictated by the physical geography (adverse climate, scarcity of food, wild beasts, etc.) they often had to face and defeat demons through prayer. In Byzantine geographical imagination eastern Mediterranean mountains were 'trial chambers for the spirit', settings for the struggle between the forces of good and evil (Schama, 1995, pp. 411–412). The paradigm had been set by Christ himself, when, at the very beginning of his ministry he was taken by Satan to an 'exceedingly high mountain' and tempted with 'all the kingdoms of the world and the glory of them' (Matthew 4: 8). Unlike other biblical peaks, the Mount of Temptation remains unidentified, though tradition mapped it in the Judean desert and marked it with a Greek-Orthodox monastery, today easily reachable from Jericho by cable car (Figure 7.3a). Other local traditions identified it with

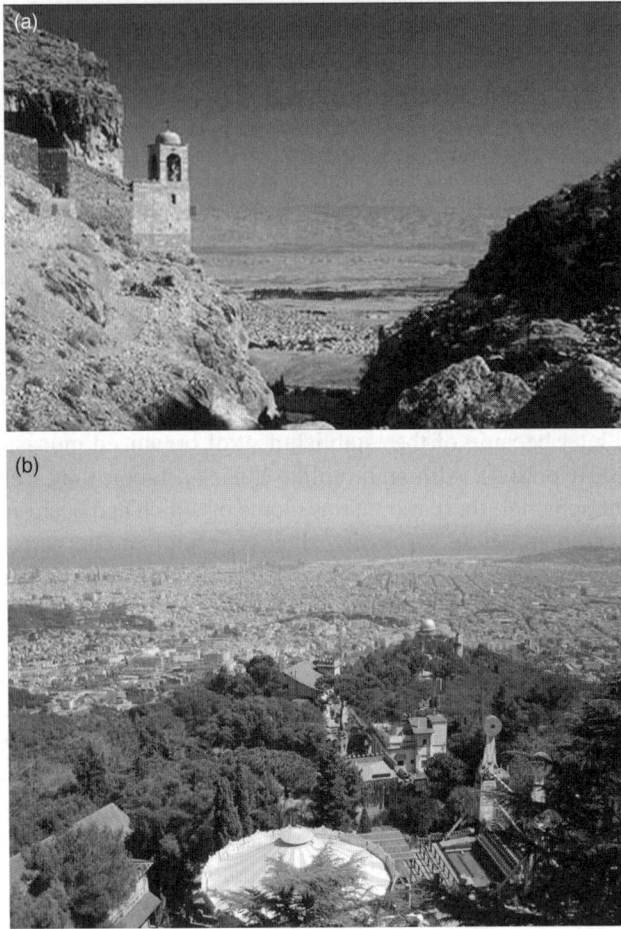

Figure 7.3   (a) View from the Mount of Temptation (source: Wikimedia). (b) View of Barcelona from Tibidabo, today topped by the Catholic church of the Sagrat Cor and an amusement park (photograph by Veronica della Dora)

geographically distant dramatic peaks around the Mediterranean: Mount Athos looming on the Aegean; Mount Saitani, a steep stony hill abruptly rising from the bare plateau of Rovòlakka in Evrytania, central Greece (Woodhouse, 1973 [1897], p. 32); and, on the other side of the Mediterranean, Mount Tibidabo the highest mountain overlooking Barcelona and named after the Latin vulgate bible verse '. . . et dixit illi haec *tibi* omnia *dabo* si cadens adoraveris me' ('And [Satan] saith unto [Christ], All these things will I give thee, if thou wilt fall down and worship me') (Figure 7.3b).

Latin monks followed their eastern counterparts in the selection of appropriate sites for their brotherhoods. In the eleventh century the most austere of the Benedictines established abbeys like Montserrat, nested at 1236 m asl, the highest point of the Catalan lowlands, or the Grande Chartreuse in the French Alps north of Grenoble, situated in a deep gorge covered by woods, behind 'bastions of inaccessible rock' (Schama, 1995, p. 414). In the Western (as in the Eastern) geographical imagination, mountains endured as *loci horridi* at least until the end of the seventeenth century and their Romantic appreciation as sublime sceneries (Nicholson, 1997 [1959]). During the Middle Ages, the mountains on the northern shore of the Mediterranean were said to be infested by brigands, witches and dragons. Until the eighteenth century, stories of encounters with the latter abounded, from the Catalan Pyrenees to the Swiss Alps. For safety reasons, in the seventeenth century access to dragon-infested peaks such as Mount Pilatus, a small peak near Lucerne believed to be the burial site of Pontius Pilate, was often forbidden by local authorities (Figure 7.4a,b). In the early eighteenth century, mountain dragons changed from being signs of diabolical contamination to become objects of scientific enquiry. Notably, Johann Jacob Scheuchzer, a Zurich professor of physics, produced a catalogue of Alpine specimens. The best ones, he wrote, 'were to be found in the sparsely

Figure 7.4   (a) Mount Pilatus, Switzerland (source: Wikimedia). (b) 'Dragon of Mons Pilatus' by Johann J. Scheuchzer, in *Itinera per Helvetiae Alpines*, 1702–11 (British Library)

inhabited cantons of the Grisons: that land is so mountainous and well provided with caves that it would be odd not to find dragons there' (Bernbaum, 1997, p. 123). Theophanic sites and *loci horridi,* Mediterranean high places, however, were not only populated by the supernatural. They have long hosted human communities too. And it is to these that we shall turn in the next section.

7.4 'Traditional' mountains

Mountainous Mediterranean stretches have always represented areas of human and natural reserves for the inhabitants of the region at large. Mediterranean mountain geographies, specifically, seem to have followed a pattern of ebb and flow, shrinking in significance during periods of peace and prosperity and growing in significance during periods of strife and scarcity. 'Mountain life, exported in generous quantities, has contributed to the overall history of the [Mediterranean] sea. It may even have shaped the origins of that history, for mountain life seems to have been the first kind of life in the Mediterranean' (Braudel, 1995, p. 52). Nonetheless, the history of the ancient Mediterranean world allows for an intricate interrelation between socioeconomic growth and cultural enrichment on the one hand, and seafaring and maritime expansion on the other. In the ancient world, where land resources became insufficient, usually maritime exchange thrived, consequently leading to a certain cultural openness and cosmopolitanism. According to Lowenthal, 'Mediterranean folk tended to be urbane cosmopolites' (Lowenthal, 2008, p. 371). The great upheavals and conquests of Mediterranean history, nevertheless, encouraged mountain settlement. Conditions of insecurity following the breakdown of Roman control in the western Mediterranean and Byzantine authority in the east made self-defence a necessity and hence mountains more appealing (McNeill, 1992, p. 91).

For example, the coastal landscape of the Aegean Sea owes its present appearance largely to fourteenth-century raids by Saracens and other pirates, forcing populations to withdraw to the interior and fortify themselves behind hilltop or mountaintop castle villages, away from the shores (Figure 7.5a,b). Such castles or castle-like villages dot the entire Mediterranean coast and have become an inextricable part of its landscape. These retreats, however, acquired much more pronounced dimensions in times of occupation by external foes, as in the case of the Ottoman occupation of the Byzantine Empire. Whereas seaborne interaction, before the fifteenth century, was the *sine qua non* of Mediterranean prosperity, it subsequently came to impoverish its peoples, giving way to segregation, expulsions and rigid distinctions (Lowenthal, 2008, p. 372), ultimately leading to a turn towards mountain life. In pre-modern Greece, the mountain landscape thus assumes centre stage in people's landscape conscience, relative to lowland or coastal landscapes. During the Ottoman occupation, 'the mountains were seen by the occupied Greeks as the Promised Land: in a nutshell, they offered a chance of freedom and the good life . . . The mountains are a Land of Cockaigne running with milk and honey, . . . but they are also the only part of the enslaved fatherland where bravery can

Figure 7.5 (a) Castle village of Anavatos, Chios island, Greece, and (b) view from Zagorochoria village, NW Pindos (photographs by Theano S. Terkenli)

flourish . . . Most importantly, the mountains are seen as the only place where the otherwise incurable taint of slavery can be healed . . . If the highlands signify freedom, the good life, anarchy and glory, the lowlands stand for slavery, grinding labour, poverty and humiliation – associations familiar to other cultures than just the Greek' (Stathatos, 1996, p. 22).

In contrast, in current times of prosperity, Mediterranean mountains have come to constitute a large proportion of the EU's 'Less Favoured Areas' in terms of general socioeconomic development. Conditions and characteristics of underde-velopment vary widely across the European Mediterranean region. Nonetheless, according to EU policy, Less Favoured Areas are generally considered either marginal/peripheral – isolated, and thus problematic – areas, or simply areas with a development deficit, either in terms of economic productivity (including un-favourable terms of competition and supply-demand correlations) or in terms of more general resource availability (Sophoulis and Spilanis, 1993; Spilanis *et al.*, 2004). These resources, in the case of Mediterranean mountainous areas, may encompass environmental/natural resources (often susceptible to technological or physical inaccessibility) or human resources (depopulation or cultural peripherality and isolation).

As already mentioned, the Mediterranean mountains, as in most mountain cases around the world – at least of the developed world – represent some of the last re-treats from the advance and spread of newer ways of life from the more populated and urbanized lowlands. Thus, they represent 'other' places, of 'other' times and 'other' cultures, increasingly threatened by extinction. Before the Second World War, the mountains of Greece still preserved ways of life inherited from the long Ottoman occupation. Whereas plains and cities were dominated and governed by the Turks and their Greek and other collaborators and allies, mountains remained hotbeds of Greek insurgence and political resistance. According to Elefantis (2002), these populations had developed thriving markets and ways of life, based on agri-culture and animal husbandry, as well as wood, masonry and wool industries. Very specific cultural systems grew out of these activities and ways of life, such as stone and wood architecture, nomadism and brigandage. In fact, this is where nomads started settling; where education and scholarship flourished in their schools; where their unique local gastronomy revolved around pies, meat, rusks and wild greens; where typical song and dance, as well as endogamy patterns, dominated (Elefan-tis, 2002). As already mentioned, for these few communities holed up in natural mountain fortresses – such as the Mani, the Souli and the Agrafa – life, however hard, poor and compromised it might be, nevertheless allowed for and symbolized a precious modicum of liberty (Stathatos, 1996, p. 23).

This mountain *modus vivendi* and widespread imaginary changed only gradu-ally through the political upheaval of the Civil War (1944–1949). The major part of transformation of mountain ways of life into a simple extension of city life occurred with the definitive linkage of mountain areas with the country's larger road system, in the 1950s (Louloudis *et al.*, 2004, p. 237; Sotiropoulou, 2007, p. 51). This trans-formation represents the irrevocable end of an era and its particular ways of life, moulded through the 'longue durée' of Greek mountain history (Braudel, 1995) – a history replicated in various locations around the Mediterranean (Pungetti *et al.*, 2008; Rackham, 2008, p. 58).

One vestige of these older Greek ways of life may be found in the patterns of animal husbandry, most of which – now of extensive and near-organic form – is concentrated in mountainous and 'less favoured areas' (75%) (Louloudis *et al.*, 2004). Across the Mediterranean, mountain people's lives bear a series of observed or imagined resemblances, such as deeply rooted community traditions that have grown out of the age-old need of mutual help, through their knowledge of nature, fauna, and often love of singing or walking; and shared modes and methods of production, such as dairy farming (Debarbieux, 2008). This specificity and vulnerability of the latter economic activity in Greece has been recognized and partly fostered by European funds – through the Common Agricultural Policy – continuing to produce high-quality products targeted towards a thriving market (Louloudis *et al.*, 2004). Nonetheless, despite all the human effort that goes into such production, in the context of a restructured livestock economy, animal husbandry in Greece remains for the most part inadequate, impoverished and parochial from every point of view (Louloudis *et al.*, 2004). In sum, necessary and imperative scientific, institutional and practical interventions towards the rectification of such multilayered problems of these mountain industries would not only upgrade them functionally and economically. Such interventions would also promote and safeguard the model – much sought by the EU – of multifunctional agriculture and animal husbandry in mountain landscapes of 'High Nature Conservation Value' farmland (Louloudis *et al.*, 2004, p. 245).

Many of the unfavourable characteristics of Mediterranean mountain areas may be remedied and rectified – and this process has already started, often to the benefit of local communities. Among natural/environmental resources, the mountain landscape itself holds a highly priced and prominent place. The stunning beauty of mountainous Mediterranean landscapes or cliffy coasts is largely the result of tectonic forces, and the unique geographical location and palaeogeographical history of the region. Generally speaking, landscape preservation and management is of paramount significance to various aspects and functions of Mediterranean mountain communities. 'Wild' nature and other environmental resources, cultural heritage of every form and genre and landscapes of high symbolic/aesthetic value have become much sought-after goods, indeed often valuable commodities, in the context of a tertiary economic sector boom, the growth of the leisure and recreation industry and the pursuit of a better quality of life. From a contemporary perspective, these mountains have represented 'traditional' enclaves and ways of life. Advances in technology (i.e. energy production alternatives and information/communication networks) and new institutional contexts (the restructuring of policy-making and international legislative, regulatory or advisory bodies) have contributed greatly towards these pursuits and opportunities. It is to these that we now turn. In the final section of the chapter, 'tradition' turns to commodity, through a rediscovery of the Mediterranean mountains by new thriving markets of various sorts, predominantly through tourism.

7.5 Commodified mountains

Mountain tourism (Sotiropoulou, 2007), geological tourism (Zouros, 2007) and ecotourism are just some examples of 'alternative' forms of tourism, which have grown exponentially in the past 20 years or more, based on the auspicious encounter of rising niche tourism demands with favourable conditions from the supply side (Coccossis and Tsartas, 2001; Tsartas *et al.,* 2004). Such favourable conditions, in our case, are provided by 'less favoured areas', such as Mediterranean mountains, offering possibilities for the consumption of 'adventure', 'wildlife', 'nature', 'difference', 'authenticity', 'history', 'isolation', etc. Assisted by improved transportation amenities and communication network capacities, increased leisure and recreation options, better marketing and promotion strategies and new trends in environmentally and ethically correct tourism (Tsartas *et al.,* 2004), 'less favoured areas', including Mediterranean mountain destinations, are enjoying a marked rise in demand for alternative tourism (Figure 7.6a,b). 'As the land is increasingly seen

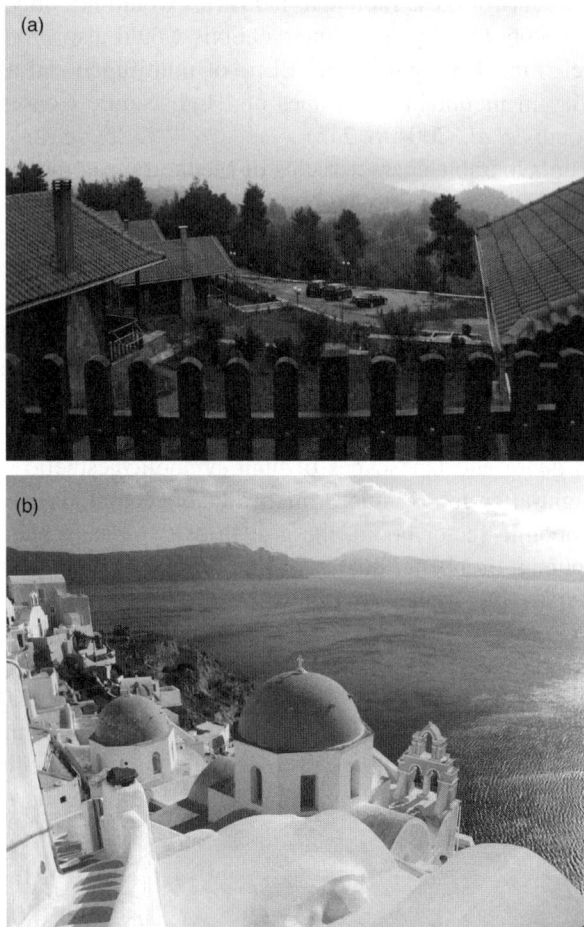

Figure 7.6   (a) Modern cottages in Pertouli, and (b) church domes overlooking caldera in Oia, Santorini (photographs by Theano S. Terkenli). (*A full colour version of this figure appears in the colour plate section*)

as a consumption resource', Denis Cosgrove noted (Sooväli-Sepping 2010:153), 'we do not want to preserve landscapes necessarily because we want to preserve some kind of authentic peasantry in the way that was the case in the nineteenth century. We want to preserve landscape because we enjoy being tourists in it, we enjoy walking and hiking in it. That is why Europeans want to preserve landscape – it is an object of consumption' (H. Sooväli-Sepping, 2007, unpublished interview).

Hence, there has been a growing demand for traffic-free mountain tracks and footpaths, permitting the enjoyment of rewarding, interesting, stimulating and/or special-interest activities (e.g. birdwatching). Such activities have inevitably contributed to landscape-related actions in the context of local policy initiatives, for example to conserve, signpost, clean and repair old paved footpaths (LEADER projects). Although few and sporadic, these initiatives lead to a more widespread landscape awareness and promotion of mountain landscapes, as well as directly contributing to the preservation of 'traditional' landscape forms, functions and meanings/values for both locals and tourists. 'Policy initiatives are only recently beginning to emerge, in response to growing concern about the future of deteriorating landscapes . . . The main lesson about the footpaths' example is that, once old landscape elements acquire new functions (due to new values regarding the landscape or activities in it), they have better chances of preservation' (Kizos *et al.*, 2007).

Presently, 'it is transalpine celebrants who, having previously chronicled Mediterranean "degeneration," now romanticize Mediterranean endurance' (Lowenthal, 2008, p. 385). Mediterranean folk emulate their forebears' endurance: 'our culture survives by allowing itself to be overwhelmed . . . We corrupted the Napoleonic invaders, and we'll deal with tourism as well. Adaptation . . . makes us unconquerable' (interview with a local, in Lowenthal, 2008). Nonetheless, it is easy to exceed Mediterranean mountain sites' carrying capacity, when winter resort infrastructure and recreation activity overwhelm the natural mountain setting. For instance, 'these days, the noble aim of "democratising skiing" seems to apply less and less . . . The ideals of the first mountain pioneers, who strove to create the best possible conditions for their leisure time in the snow, has been reduced to . . . selling the winter sports "products" at all costs' (Venema, 2008, p. 124). The unrelenting advance of winter sport tourism has therefore 'not only pushed the much older forms of mountain tourism and leisure pursuits, the true forms of Alpinism such as climbing and hiking, into the background, it is in fact at variance with it. The heavy infrastructure, construction and spoiling of the landscape has undermined the conditions or these forms of summer recreation' (Venema, 2008, p. 126).

There is another danger lurking in the background of the various contemporary and fast-advancing alternative forms of tourism encountered in Mediterranean mountain settings and communities. 'Nature-tourism' or ecotourism growth tends to idealize not only traditional modes of production, but also an entire rural past – often presented as homogenized and ahistorical – as a condition of stasis and a type of balance with nature, in contrast to the industrial way of life (Deltsou in Galani-Moutafi, 2002). 'For achieving a continuity of the balance between nature and culture, an outlet is provided through the establishment of protected natural areas, which – along with their residents – represent the past and the state of isolation. . . . Therefore, ecotourism contributes towards the shaping of a particular perception concerning both the past and the people's present' (Galani-Moutafi, 2002, p. 274).

However, mountain commodification may proceed along more benign and sustainable lines. One such example is energy production, which, in some cases, may imply the return to older technological solutions to longstanding problems, as in the form of watermills (Liarikos, 2004). Another example is the development of local industries, through agri-environmental measures enhancing landscape multifunctionality and preservation (i.e. through flora and fauna conservation), while creating and diversifying local employment – including immigrant labour – and producing high-quality local products (Louloudis *et al.*, 2004) in rising demand by niche economies (dairy products, wine, forest products, etc.). However, no matter what commodities, amenities or services these new mountain tourism industries are based on (winter sports, museums of industrial history, agrotourism, geoparks, etc.), they run the risk of simply sapping the destinations of their resources, leaving them in a condition of environmental degradation, sociocultural deterioration and economic impoverishment and dependence. In order to prevent such repercussions, paramount are circulation and recycling of tourism benefits and profits back into local industry and sustainable development; preservation and quality control of those resources that attracted tourism in the first place, as well as of the whole tourism setting itself; and – last, but not least – long-term tourism planning, encouraging local cooperation initiatives, a broad/democratic basis of decision-making and proper management of the operation, resources and impacts of tourism on the local community, culture and environment. In this way, instead of rampant, laissez-faire tourist development and place consumption and degradation, local mountain communities may be assured and reassured not only of their place identity, quality of life, cultural promotion and landscape preservation, but also of sustainable, equitable and ecologically safe economic development.

7.6 Conclusion

Mountains have never failed to stir human imagination. The most prominent and seemingly 'durable' of geographical objects, they inspire paradoxical responses: awe and fear, attraction and repulsion, security and remoteness. They are familiar landmarks that remain insistently 'other'. A plethora of narratives, practices and moral attributes has been layered over mountain rocks – variously in different cultural traditions. As this chapter has shown, Mediterranean mountains represent especially rich palimpsests, as 'almost every place in the Mediterranean world has at one time or another been pagan, Christian and Muslim' (Horden and Purcell, 2001, p. 403), yet they are also at the same time fragile ecosystems.

Despite its coastal orientation the Mediterranean remains 'a region of mountains, of complex and fragmented relief' (King, 1997, p. 8). Through the centuries, Mediterranean mountain geographies attracted pagan priests, biblical prophets and

Christian hermits alike. They also made them bulwarks for religious and political resistance, as well as reservoirs of past customs, of traditional *genres de vie*. As such,
Mediterranean high places are becoming valuable resources for tourist consumption; objects of nostalgia for lost natural 'otherness' and a lost past 'otherness' –
in other words, for authenticity. But nonetheless, they remain places of limited resources and high ecological fragility. Since the beginning of the twentieth century,
their residents have been abandoning mountains for lowland cities and villages, although at different rates and time periods around the Mediterranean Basin.

Nowadays, however, we are witnessing a reverse – albeit partial – flow back to
mountain communities, to the rural family hearth, to the ancestral home, to old
villages slowly turning into either lucrative winter resorts or thriving, quaint, 'traditional' second-home communities for well-off urbanites. There exist, of course, a
series of symbolic Mediterranean mountain geographies, ranging from cultural images for tourist consumption to problem-ridden peripheries for the local populations
to national or family hearths for the rest of the population, constructed in the col-
lective imagination with an orientation towards a historical – or ahistorical, for that
matter – past. According to this latter imaginary, during most of the year, mountains
are perceived as essentially uninhabited landscapes, while during holidays and especially during the summer, they come alive again to cater to the recreational needs
of urbanites and other tourists. For all of the above reasons, these landscapes need to
be preserved and managed in sustainable ways that are appropriate, both temporally
and spatially, to the needs of modern society, without compromising the natural and
cultural inheritance that makes up the uniqueness of contemporary Mediterranean
mountain geographies.

References

References marked as bold are key references.

Bernbaum, E. (1997) *Sacred Mountains of the World.* **Berkeley: University of California
Press.**
Braudel, F. (1995) *The Mediterranean and the Mediterranean World in the Age of Philip II,*
vol. 1. Berkley and Los Angeles: University of California Press.
Braudel, F. (2000) La terra. In: Braudel, F. (ed.), *Il Mediterraneo: lo spazio, la storia, gli uomini,
le tradizioni.* Milano: Bompiani, pp. 11–30.
Coccossis, H. and Tsartas, P. (2001) Sustainable Tourism Development and the Environment [in
Greek]. Athens: Kritiki.
Cosgrove, D. and della Dora, V. (eds) (2009) *High Places: Cultural Geographies of Mountains,
Ice and Science.* **London: IB Tauris.**
Debarbieux, B. (1998) The mountain in the city: social uses and transformations of a natural
landform in urban space. *Ecumene* 5:399–431.
Debarbieux, B. (2004) The symbolic order of objects and the frame of geographical action: an
analysis of the modes and the effects of categorisation of the geographical world as applied to
the mountains in the west. *GeoJournal* 60:397–405.
Debarbieux, B. (2008) Linking mountain identities throughout the world: the experiences of
Swiss communities. *Cultural Geographies* 15:497–517.

della Dora, V. (2005) Alexander the Great's mountain. *Geographical Review* 95:489–516.

Elefantis, A. (2002) The mountains: in search of a tragic life [in Greek]. *O Politis* 98:36–38.

Eliade, M. (1959) *The Sacred and the Profane: The Nature of Religion.* New York: Harcourt, Brace and World, Inc.

Galani-Moutafi, V. (2002) *Research on Tourism in Greece and Cyprus: An Anthropological Perspective* [in Greek]. Athens: Propombos.

Greenfield, R. (2000) *The Life of Lazaros of Mt. Galesion: An Eleventh-century Pillar Saint.* Washington, DC: Dumbarton Oaks.

Horden, P. and Purcell, N. (2001) *The Corrupting Sea.* Oxford: Blackwell.

Kadas, S. (1998) *Mount Athos: The Monasteries and Their Treasures.* Athens: Ekdotike Athenon.

King, R. (1997) Introduction: an essay on Mediterraneanism. In: King, R., Proudfoot, L. and Smith, B. (eds), *The Mediterranean: Environment and Society.* London: Arnold, pp. 1–11.

Kizos, T., Spilanis, I. and Koulouri, M. (2007) The Aegean islands: a paradise lost? Tourism as a driver for changing landscapes. In: Pedroli, B., van Doorn, A., de Blust, G., Paracchini, M.L., Wascher, D. and Bunce, F. (eds), *Europe's Living Landscapes.* Wageningen: KNNV Publishing, Zeist, pp. 333–348.

Liarikos, C. (2004) Alternative forms of tourism and mountain development: evidence from the outdoor museum of hydromovement, in Dimitsana, Greece [in Greek]. In: Spilanis, I., Iosifides, T. and Kizos, A. (eds), *Development Strategies in Less Favoured Areas.* Athens: Gutenberg, pp. 307–328.

Louloudis, L., Vlachos, G. and Theocharopoulos, I. (2004) The dynamic of local survival in Greek less favoured areas [in Greek]. In: Spilanis, I., Iosifides, T. and Kizos, A. (eds), *Development Strategies in Less Favoured Areas.* Athens: Gutenberg, 235–66.

Lowenthal, D. (2008) Mediterranean heritage: ancient marvel, modern millstone. *Nations and Nationalism* 14:369–392.

Mavian, L. (1992) Ruolo della mitologia nella percezione della natura e nell'organizzazione delle sue risorse: luoghi mitici o illustri. In: Luginbuhl, Y. (ed.), *Paesaggio mediterraneo.* Milano: Electa, pp. 36–41.

McNeill, J.R. (1992) *The Mountains of the Mediterranean World: An Environmental History.* Cambridge: Cambridge University Press.

Nes, S. (2007) *The Uncreated Light: An Iconographical Study of the Transfiguration in the Eastern Church.* Grand Rapids, MI, and Cambridge: Eerdmans.

Nicholson, M.H. (1997) [1959] *Mountain Gloom and Mountain Glory.* Seattle and London: University of Washington Press.

Norberg-Schultz, C. (1979) *Genius Loci: Towards a Phenomenology of Architecture.* London: Academy Editions.

Pattenden, P. (1983) The Byzantine early warning system. *Byzantion* 53:258–299.

Peatfield, A. (1983) The topography of Minoan peak sanctuaries. *Annual of the British School of Athens* 78:273–279.

Pungetti, G., Marini, A. and Vogiatzakis, I.N. (2008) Sardinia. In: Vogiatzakis, I.N., Pungetti, G. and Mannion, A.M. (eds), *Mediterranean Island Landscapes: Natural and Cultural Approaches.* Dordrecht: Springer, pp. 143–169.

Rackham, O. and Moody, J. (1996) *The Making of the Cretan Landscape.* Manchester: Manchester University Press.

Rackham, O. (2008) Holocene history of Mediterranean island landscapes. In: Vogiatzakis, I.N., Pungetti, G. and Mannion, A.M. (eds), *Mediterranean Island Landscapes: Natural and Cultural Approaches.* Dordrecht: Springer, pp. 36–60.

Rutkowski, B. (1986) *Cult Places of the Aegean*. New Haven and London: Yale University Press.

Schama, S. (1995) *Landscape and Memory*. New York: Vintage Books.

Scully, V. (1979) *The Earth, the Temple and the Gods: Greek Sacred Architecture*. New Haven and London: Yale University Press.

Semple, E.C. (1932) *The Geography of the Mediterranean Region: Its Relation to Ancient History*. London: Constable.

Sooväli-Sepping, H. (2010) The Role of Geography in the Twenty-First Century: Interview with Denis Cosgrove. In: Dora, V.D., Digby, S. and Basdas, B. (eds), *Visual and Historical Geographies: Essays in Honour of Denis E. Cosgrove*. Historical Geography Research Group Series, London: Royal Geographical Society, pp. 147–153.

Sophoulis, C. and Spilanis, I. (1993) Pour une stratégie de développement insulaire. *Revue de l'Économie Régionale* 41:33–44.

Sotiropoulou, E.C. (2007) New development dimensions and perspectives in contemporary agricultural space in Greece: a study of three villages in less favoured areas [in Greek]. In: Kizos, T., Iosifides, T. and Spilanis, I. (eds), *Special Development Issues in Less Favoured Areas*. Athens: Gutenberg, pp. 36–68.

Spilanis, I., Kizos, A. and Iosifides, T. (2004) Economic, social and environmental dimensions of development in less favored areas (L.F.A.s) [in Greek]. In: Spilanis, I., Iosifides, T. and Kizos, A. (eds), *Development Strategies in Less Favored Areas*. Athens: Gutenberg, pp. 13–38.

Stathatos, J. (1996) *The Invention of Landscape: Greek Landscape and Greek Photography 1870–1995*. Thessaloniki: Camera Obscura.

Talbot, A.M. (2001) Les saintes montagnes à Byzance. In: Kaplan, M. (ed.), *Le sacré et son inscription dans l'espace à Byzance et en Occident*. Paris: Byzantina Sorbonensia, pp. 263–275.

Tsartas, P., Stogianidou, M. and Stavrinoudis, T. (2004) Less Favored Areas as tourist destinations: issues of organization and management [in Greek]. In: Spilanis, I., Iosifides, T. and Kizos, A. (eds), *Development Strategies in Less Favored Areas*. Athens: Gutenberg, pp. 295–306.

Venema, H. (2008) Mountain. In: Hazendonc, N., Hendricks, M. and Venema, H. (eds), *Greetings from Europe: Landscape and Leisure*. Rotterdam: OIO Publishers, pp. 118–129.

Williams, M. (1989) *Landscape in the Argonautica of Apollonius Rhodius*. Frankfurt and New York: Peter Lang.

Woodhouse, W.J. (1973) *Aetolia: Its Geography and Topography*. New York: Arno Press.

Zouros, N. (2007) Promotion and protection of natural monuments and local development in rural mountainous and island areas [in Greek]. In: Kizos, T., Iosifides, T. and Spilanis, I. (eds), *Special Development Issues in Less Favoured Areas*. Athens: Gutenberg, pp. 179–200.

8

Land use changes

Vasilios P. Papanastasis

8.1 Introduction

Land is an important natural resource for the survival and prosperity of people and for the maintenance of terrestrial ecosystems, including the Mediterranean mountains. It is often regarded as equivalent to soils and topography but this is not correct. According to the FAO (1997), land has a wider sense that implies not only the land surface and its underlying superficial deposits but also its attributes such as climate, water, plants, animals and humans with their activities. This means that land is a complex entity encompassing physical, biological, environmental, infrastructural, social and economic factors that interact with each other.

On the other hand, land is finite and immovable, which suggests that it can neither increase in size nor be transported. On the contrary, people who use the land for the production of commodities useful for their survival may move and increase in numbers and impose high demands, or decrease and reduce their pressure. This means that there is a constantly changing relationship between people and land (Davis, 1976). Land use therefore refers to the type of human activity taking place on land. It is characterized by the arrangements, activities and inputs by people to produce, change or maintain a certain land cover type (Di Gregorio and Jansen, 1998). This implies a direct link between land cover and the actions of people on their environment. Consequently, there is a close relationship between land use and land cover, the latter considered to be the observed biophysical cover of the Earth's surface (Cihlar and Jansen, 2001).

Mediterranean mountains have a long history of human intervention, which has modified land cover and resulted in numerous land use changes over time. In classical Greece (fourth to fifth centuries BC), for example, people exploited the environment and used or abused its natural resources such as forests and game species (Papanastasis *et al.*, 2010). However, Grove and Rackham (2001) claim that modern Mediterranean landscape has surprisingly changed little since ancient times, except for coasts, deltas and marshes. On the contrary, McNeill (1992) argues that for most of the mountains the changes that destroyed the environment and left behind

Mediterranean Mountain Environments, First Edition. Edited by Ioannis N. Vogiatzakis.
© 2012 John Wiley & Sons, Ltd. Published 2012 by John Wiley & Sons, Ltd.

skeletal landscapes are comparatively recent, dating back no more than 200 years, and were greater than the slower and more modest changes of the earlier periods. Since the Second World War, with the introduction of heavy machinery such as tractors and bulldozers for the exploitation of natural resources, such changes have been devastating to Mediterranean ecosystems (Naveh and Lieberman, 1994; Papanastasis, 2004). Currently, mountain ecosystems are changing more rapidly than at any time in human history (Körner and Ohsawa, 2005).

Land use changes involve both space and time and result in modification of the Mediterranean mountains. In this chapter, the drivers as well the historical evolution of land use changes in the major mountain ranges of the Mediterranean Basin are reviewed and their impacts on mountain environments are discussed.

8.2   Drivers of land use changes

Land use changes in the Mediterranean mountains are driven by several factors, which may be sociopolitical, economic and environmental. Of the sociopolitical factors the most important is population. Mediterranean mountains have a certain carrying capacity, that is the maximum number of people whose needs they can sustain. If their population is below this limit then the impact of people on natural resources may result in only temporary perturbations and no permanent changes are made to ecosystems. In contrast, if the population has grown beyond the carrying capacity, then permanent changes occur that may lead to environmental degradation (e.g. soil erosion).

Mediterranean mountains have experienced both expansions and contractions of their population. According to McNeill (1992), mountain populations began to grow in the late eighteenth century after the introduction of maize and potato, which made the survival of people much easier. In the late nineteenth century, large-scale emigration of mountain people began, which slowed down in the early twentieth century and then accelerated again after the Second World War, largely due to the better living conditions in the lowlands and urban centres. Similar fluctuations occurred in earlier periods, mainly caused by historical events. In the Psilorites mountain of Crete, for example, the Venetian and Turkish occupations resulted in the reduction of its population, reflecting a similar decrease in the whole island (Figure 8.1). Also, when the Ottomans conquered Greece in the fifteenth century, Christians were forced to move to the mountains to avoid mixing with the Muslims, who came from Anatolia and settled in the lowlands (Vakalopoulos, 1964).

Among the economic factors, market forces may cause land use changes in the Mediterranean mountains. If mountain villages are not fully integrated with large markets in the lowlands, then there is no strong motivation for production of marketable goods and services, and consequent potential impact on natural resources. In contrast, if roads are opened and connections are established with the lowlands, goods start to be imported into the mountain villages, which motivates the exploitation of the local resources, mainly forests, for the production of marketable goods

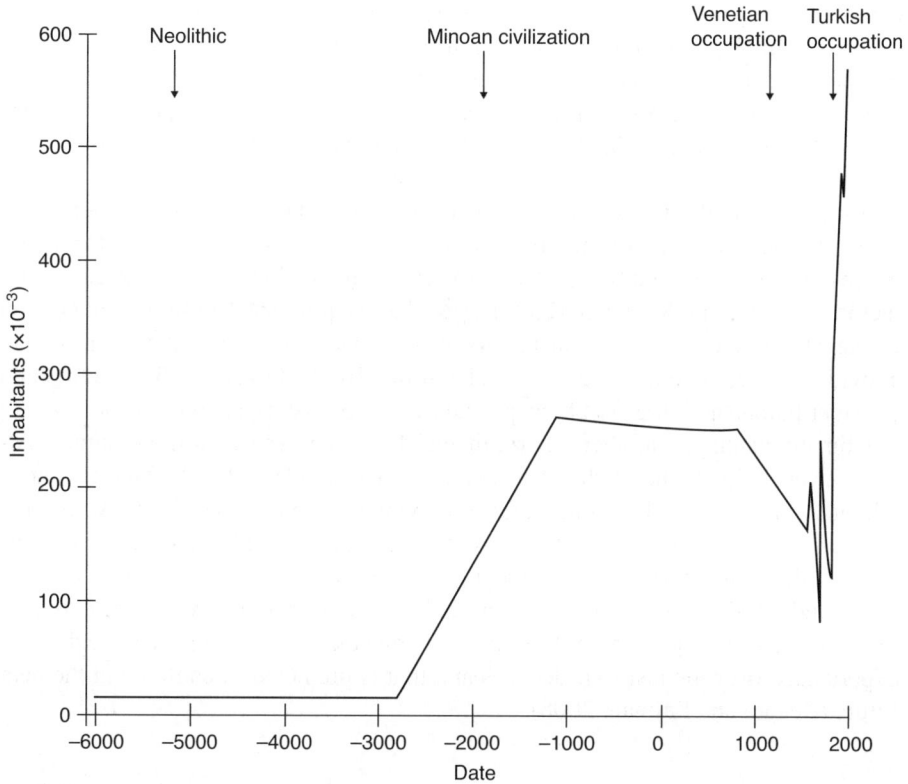

Figure 8.1 Evolution of Crete's population through the centuries. Reproduced from Lyrintzis and Papanastasis (1995), with permission

in exchange. Historically, the latter development occurred in the eighteenth and especially in the nineteenth century, when modern technology helped to improve transport from mountains to markets in the lowlands leading to massive deforestation (McNeill, 1992). A case study is the destruction of the Taurus mountain forests at the end of the nineteenth century in order to produce construction materials for the Suez Canal (Regato and Salman, 2008).

Technology is another economic factor that forces land use changes. In the past, farming with the Hesiod's plough or cutting timber with an axe or moving grazing animals on foot did not have a great impact on mountain natural resources. With mechanization after the Second World War, such as the introduction of the tractor for farming, the chainsaw for logging and vehicles for transporting animals from the lowlands to uplands and vice versa, the impact on natural resources was much stronger and more widespread. In Greece, a dense network of access roads has been built in several mountain rangelands in the last few decades in order to facilitate the movement of shepherds and their pick-up vehicles from villages to the livestock farms (Papanastasis, 2004). Such a network has had a considerable impact on the environment by breaking up habitats and resulting in severe soil erosion. Additional modern technological interventions include mountain homes, solar and wind energy generators, and small-scale hydropower plants.

Finally, among the environmental factors, geology and geomorphology are important factors in causing land use changes. Soil erosion is usually caused by human activities but it is largely associated with the nature of the parent material as well. Schist, for example, is more erosive than limestone but vegetation recovers from destruction faster on the former than on the latter due to the different soil formation processes (Conacher and Sala, 1998). However, soil erosion may be also caused by tectonics. Grove and Rackham (2001) argue that the primary determinant of erosion in the Mediterranean Europe of historical times was tectonics rather than human activities. Other scholars (e.g. van Andel *et al.,* 1990; Butzer, 2003) disagree and consider human activities to be the primary cause of historical soil erosion.

Climate change is another environmental factor causing land use changes. Although vegetation in the Mediterranean mountains has adapted to the Mediterranean climate, characterized by a cool and rainy winter period followed by a warm and dry summer period, extreme events may result in significant alterations. Such events may be a prolonged dry period (drought) or a torrential rain event leading to landslides and floods (Grove and Rackham, 2001). These extreme events are expected to become more frequent with the climate change that the Mediterranean region has experienced over the last four decades and that is predicted to continue in the near future (Regato and Salman, 2008).

8.3   Major land uses and their historical evolution

8.3.1   Wood cutting

Wood cutting has been an important land use practice for the survival of mountain people, who use wood as a construction material for their homes, as a fuel and as a source of income. However, given the inaccessibility of the Mediterranean mountains, it is unlikely that commercial logging was practised widely in the mountain forests in the past. Not until the nineteenth century, when roads, trucks, railroads, cable lifts and other devices made mountain forests accessible, did this level of exploitation become feasible (McNeill,1992).

Nevertheless, there are disastrous cases of commercial logging in the past. One such example is the destruction of the cypress forests of the mountains of Crete in the Bronze Age to provide construction material for the Minoan navy (Tsoumis, 1986). Another case is the clearing of the cedar forests of Mount Lebanon by the Phoenicians, who became famous sea traders at the end of the second millennium BC thanks to their cedar ships (Thirwood, 1981). In the classical period, Athens was short of timber for construction and it had to import it from more wooded regions of Greece (Macedonia) and elsewhere (Sicily) (Meiggs, 1982).

Compared to timber logging for construction or trading, firewood collection has been a more widespread land use practice in the Mediterranean mountains. This is because mountain people could use alternative, more reliable and durable construction materials for their homes in several regions (e.g. stones). However, they still needed firewood for cooking and heating, and its collection was practised around the villages within walking distance, with no need for roads or trucks for transportation. It can be considered as the main driver of land use changes around the villages.

In addition to firewood collection, charcoal making has also been a land use practice of mountain people for ensuring an alternative source of fuel for cooking and heating. This had a greater impact on forests than firewood gathering. Charcoal was often used for commercial purposes to meet the demand of distant markets located in the lowlands and the urban centres.

After the Second World War, the demand for firewood and charcoal in many mountain communities was reduced or even ceased altogether as populations decreased and households switched to modern sources of energy such as electricity for cooking and oil for heating.

8.3.2 Pastoralism

Pastoralism has been a dominant economic activity in the Mediterranean Basin since humans domesticated sheep, goats and cattle in the Neolithic period and used them as a source of food and fibre. Once they were introduced, livestock became part of the environment, with which they evolved together over the millennia. Their impact was clearly related to human history, being high at periods with high human populations or adverse political conditions (e.g. wars) and low at periods with low human populations or with political stability and economic prosperity.

In ancient times, livestock husbandry was closely related to arable agriculture, suggesting that animals were kept near the farms; there is no evidence for long-distance movements such as transhumance (Papanastasis et al., 2010). Later, livestock husbandry became more independent of arable agriculture and also moved to the Mediterranean mountains in order to make use of their forage resources during the summer, when the lowlands could not provide feed to the animals due to the seasonality of the Mediterranean climate. The complementary exploitation of forage resources between highlands and lowlands, known as transhumance, became a common practice in the mountains of all Mediterranean countries and resulted in a complex and widespread network of routes – called cañadas in Spain (Figure 8.2), drailles in France, fratturi in Italy and diadromoi in Greece (Ruiz and Ruiz, 1986; Ispikoudis et al., 2004; Biber, 2010). In Spain, transhumance reached its peak during the fifth to fifteenth centuries but after the Industrial Revolution it started to decline and accelerated dramatically during recent decades due to changes in socio-cultural and economic factors (Ruiz and Valero, 1990).

Figure 8.2 A drove road (cañada) in Spain with a flock of sheep moving to the mountains (courtesy of Antonio Gomez Sal)

With the increased colonization of the Mediterranean mountains from 1700 to 1900, pastoralism became an important human activity that contributed significantly to land use changes. However, its impact was different in different mountain regions. For example, the Taurus mountains of Turkey and the Pindos mountains of Greece experienced a greater influence of pastoralism than the Apennines of Italy, Sierra Nevada of Spain and the Rif mountains of Morocco because in the former nomads had conquered and controlled for longer than in the latter (McNeill, 1992).

8.3.3 Wildfires

Fire is a major ecological factor in the Mediterranean environment and has played a decisive role in post-glacial geological and cultural evolution in the Mediterranean Basin (Naveh, 1975). Humans have used it as a means to modify the environment to their benefit since time immemorial, and this has caused significant changes to Mediterranean landscapes and ecosystems (Naveh and Lieberman, 1994; Moreno *et al.,* 1998; Arianoutsou, 2001). After reviewing the history of burning from literary sources, Liacos (1973) concluded that fire was used as a tool to open up forests and convert them to grazing land since the second millennium BC. Pastoral burning was also used in Roman times (Hughes, 1983) as well as in subsequent historical periods. Fire was used not only in open grazing lands but also to suppress undesirable woody vegetation in rangelands. This practice is still used in several parts of the Mediterranean such as Corsica, Sardinia and Crete (Papanastasis, 1998).

8.3.4 Agriculture

Clearing land of natural vegetation to grow arable crops is a very old land use practice in the Mediterranean Basin. In the beginning, the impact of agriculture on the landscape was not great because human numbers were small and the changes were restricted to the lowlands. After the ancient civilizations developed, the impact of agriculture became greater but again it altered the plains far more than the uplands. The latter were infiltrated by agriculture in the medieval and early modern times, when people were forced to move to the mountains to avoid lowland malaria, while the introduction of the American crops such as maize and potato and the invention of irrigation techniques made their life easier (McNeill, 1992).

Cultivation on mountain slopes destabilizes soil because it removes all the natural plant cover and exposes soil to accelerated erosion. It is one of the most drastic human interventions in the environment. In order to ensure productivity, mountain farmers had to invent terracing to hold the soil in situ and prevent its washing down slope. Archaeologists claim that terraces have been used by Mediterranean farmers since ancient times (van Andel *et al.,* 1990; Rackham and Moody, 1992). However, with the large-scale emigration of mountain people in the late nineteenth century, all these terraces collapsed for lack of maintenance leading to severe erosion. Signs of old terraces can be seen in almost all Mediterranean mountains (Figure 8.3).

Like livestock husbandry, agriculture was not universally well developed in all Mediterranean mountain regions. Out of the five ranges he studied in detail, McNeill (1992) found that agriculture was historically more important in the Sierra Nevada of Spain, the Lucanian Apennines of Italy and the Rif mountains of Morocco than in the Taurus mountains of Turkey and the Pindos mountains of Greece.

8.3.5 Mining

Mining for extraction of minerals has been a land use practice in certain Mediterranean mountains. It is a very drastic activity that leads to much degraded landscapes. McNeill (1992) mentions the case of lead mining in the Sierra de Gador of Spain, which started in 1818 and ended in the 1880s, as well as the lead and iron mining in the Rif mountains of Morocco, which was carried out by Carthaginians and Romans in the past and by Spanish and British companies in modern times. Mining activities are still going on in several Mediterranean mountains. Such a case is the bauxite mining in Parnassos and Giona mountains of Greece, which has denuded the landscape (Figure 8.4).

Figure 8.3   Abandoned old terraces in the White mountains of Crete colonized by phryganic vegetation (photo by the author). (*A full colour version of this figure appears in the colour plate section*)

Figure 8.4   A landscape denuded by bauxite mining on Giona Mountain, central Greece (photo by the author). (*A full colour version of this figure appears in the colour plate section*)

8.4 Recent land use changes

8.4.1 Morocco

The reduction of forest cover has been the most important land use change in Moroccan mountains over the last few years. Rejdali (2004) reports that Morocco is losing an estimated 30 000 ha/year of forest cover due to the high demand for forest products by an increasing rural population, the expansion of ploughing to create more arable lands and the uncontrolled and intensive grazing by livestock, which are forced to graze in more confined areas following the intensive cultivation. In the central High Atlas mountains, in particular, forests were reduced by 20.7% between 1976 and 1996 resulting in a significant increase of the erosion potential (Merzouk and Dhman, 1998).

There has also been a significant change in the traditional agricultural system in the Moroccan mountains. In the past, this system was based on sedentary communities, which depended on two complementary and integrated land use types: intensive cropping on irrigated terraces and livestock husbandry, mainly sheep and goats. Since the 1970s this equilibrium has been broken due to the very rapid population growth, which has put great pressure on the land leading to degradation of the natural vegetation (Bencherifa, 1983). In addition, the traditional pastoral nomadic migrations have ceased and the permanently settled people are engaged in new agricultural activities (Bencherifa and Johnson, 1991).

8.4.2 Portugal

Recent land use changes in the Portuguese mountains were initiated in the 1960s and were largely dictated by the gradual emigration of the rural inhabitants, mainly the active generation, to the urban centres and the subsequent reduction of agricultural activity. In a diachronic study of land use changes between 1947 and 1990 carried out in the Sierra de Malcata of central east Portugal, Cohelho-Silva et al. (2004) found that the area covered by permanent crops such as olive groves had diminished; the area of temporary crops such as cereals reached a maximum in 1958 but thereafter started to decrease; and shrublands and forests had increased since 1958. At the same time, the resident human population decreased by 50% and farm animals declined in numbers by 54%. Similar results were also recorded in the Sierra de Monchique of southern Portugal, where a diachronic study of land use changes between 1966 and 1996 showed a total abandonment of rainfed agriculture on slopes and its replacement with matorral (Krohmer and Deil, 2003).

In northern Portugal, forest plantations created a new land use type. In the National Park of Alvão, for example, pine plantations were established in the 1950s at the expense of rangelands, thereby significantly altering the landscape (Timóteo et al., 2004). By contrast, in a mountain landscape of northeast Portugal, annual crops decreased by 43.5% from 1979 to 2002 due to the loss of economic competitiveness and the decline and ageing of the rural population (Pôcas *et al.*, 2011a).

Land abandonment and human-driven changes such as pine plantations have increased the fire risk and the incidence of large and devastating wildfires. After studying these wildfires in central and northern Portugal over a period of 13–15 years from 1990, Silva *et al.* (2011) concluded that fire has a determinant role in land use changes of the studied area, as it favoured shrubland persistency and the conversion of other land use types to shrublands and forests.

8.4.3 Spain

Spanish Mediterranean mountains have become a marginal territory since the 1950s when the rural population started to emigrate to the big cities in the lowlands resulting in farmland abandonment and loss of livestock farmers. In the Camero Viejo of northwestern Spain, Lasanta *et al.* (2011) have found that the cultivated area was reduced by 99% between 1956 and 1995, with the abandoned fields representing 99.4% of the cultivated area and 40.4% of the entire study area. The terrace walls collapsed in 1995 leading to severe soil erosion. Also, the number of sheep fell dramatically by 1995, and they were largely replaced by free-grazing cattle of imported breeds that tend to overgraze the abandoned fields resulting in significant soil erosion. Finally, the human population declined by 86.2%.

In the Spanish Central Pyrenees, the most significant land use changes since the 1950s have been the decline of sheep and the expansion of cattle breeding; the reduction of the cultivated area; the decline of cereal production in favour of pasture; and the reforestation of many abandoned hill slopes (Garcia-Ruiz *et al.*, 1996). In addition to reforestation, the abandoned lands were also invaded by spontaneous forest vegetation. Reduction of the farming area occurred simultaneously with population decline (Lasanta-Martinez *et al.*, 2005). Also in the Central Pre-Pyrenees, an expansion of the forested area was recorded between 1957 and 1996. This was caused by spontaneous afforestation of abandoned fields and an increase of the canopy density of the existing forests (Poyatos *et al.*, 2003). The same processes, namely the expansion of the forest area and the increasing canopy density of forests already present in 1957, occurred also in the National Park of the Pyrenees (Gracia *et al.*, 2011). Ameztegni *et al.* (2010) found that land use change played a more important role than climate in the increase of forest cover.

A new land use in the central Spanish Pyrenees is tourism, and especially skiing. However, the area influenced by the ski resorts is restricted to nearby municipalities, which show positive demographic changes and a continuation of the primary activities (Lasanta *et al.*, 2007).

In southeast Spain, with its drier climate, cropland abandonment resulted in an increase of alfa (*Stipa tenacissima*, or 'esparto') grasslands while shrublands decreased because of the extensive pine afforestation programme that was initiated in 1996 (Bonet *et al.*, 2004). A reduction of the area covered by dehesa woodlands in southwestern Spain was also documented between 1950 to 1990, due to the reduction of tree cover in favour of intensive agriculture and the almost complete destruction of the traditional silvopastoral system (Regato-Pajares *et al.*, 2004).

In the mountain environment of the Sierra Nevada of Andalucía, rural depopulation of almost 50% since the 1950s has resulted in the desertification and sometimes abandonment of irrigated terraces due to labour shortages. These abandoned terraces have been converted to grassland or been invaded by matorral and turned over to grazing (Douglas *et al.*, 1996).

8.4.4 Southern France

In Mediterranean France, land use changes started during the nineteenth century, but land abandonment began and increased in the twentieth century, until it became pronounced with the rural depopulation of the 1960s (Tatoni *et al.*, 2004). In the Madrès mountain area of the eastern Pyrenees, grasslands were reduced by 73% while the forest cover doubled in size between 1953 and 2000 (Roura-Pasqual *et al.*, 2005). Also, in the southern French Pre-Alps, the period between 1956 and 1991 was characterized by a marked extension of forest and an increase in shrub areas at the expense of ploughed lands. These changes were primarily prompted by the human population decline as well as by the reduction of sheep and goat numbers, and secondarily by environmental factors such as the nature of bedrock, elevation and slope (Taillefumier and Piégan, 2003). Finally, in the Malay Massif of the same mountains, forests covered only 56% of the surface in the mid-nineteenth century but since then Scots pine (*Pinus silvestris*) has colonized the whole mountain following the gradual abandonment of agricultural fields and the cessation of grazing pressure, especially after the 1960s (Chauchard *et al.*, 2007). A similar transformation also occurred in the interior of Corsica, where 70% of the land surface was occupied by cereal production on terraces in the nineteenth century, but nowadays these terraces are fully covered by maquis (Etienne *et al.*, 1998).

8.4.5 Italy

The abandonment of agricultural land and traditional cultivation practices due to rural emigration is a common and widespread phenomenon in the Mediterranean mountains of Italy. As a result, forest cover has increased over the mountains (Falcucci *et al.*, 2007). In the northern Apennines (Tuscany), for example, 24% of the agricultural land was transformed to shrubland between 1947 and 1993, parts of which eventually were transformed to woodlands (Torta, 2004). The same afforestation process was also observed in the province of Siena, where the recovery of forest was at the detriment of semi-natural and agricultural areas (Geri *et al.*, 2010a). Also, in the western Italian Alps, heavy grazing and tree cutting prevented stone pine (*Pinus pinea*) establishment until 1960, but since then stone pine has regenerated massively due to the cessation of grazing and tree felling (Motta *et al.*, 2006).

In southern Italy, the abandonment of both cultivated areas and forest lands in the mountains resulted in the concurrent urbanization and intensification of agriculture in the lowlands and coastal regions between 1959 and 1984. In the Amalfi peninsula of the west coast as well as in the Molise region of the east coast, forest cover greatly increased between 1954 and 1974 (Mazzoleni *et al.*, 2004a). In the Molise region, transhumance of sheep and goats was an important activity in the past but over recent decades it has dramatically reduced (Mazzoleni *et al.*, 2004a; Susmel *et al.*, 2004). In contrast, no significant land use changes occurred in the Iblei mountains of Sicily between 1856 and 1990 suggesting that the landscape remained relatively stable over this period (Di Pasquale *et al.*, 2004).

8.4.6 Greece

Recent land use changes in the Greek mountains were initiated in the 1950s and especially in the 1960s when the rural population, mainly its active members, started to emigrate to the urban centres and abroad. Between 1961 and 2000, arable land in the mountains was reduced by 27% (Papanastasis, 2007). For livestock grazing, there was a decline of transhumance. Numbers of sheep and goats, the animals that traditionally underwent this ancient practice, sharply decreased (by 40%) between 1961 and 1971, as the ethnic groups involved (e.g. the Saracatsani and Vlachs) became permanently settled in the lowlands (Ispikoudis *et al.*, 2004).

In the Portaikos-Pertouli valley of the central Pindos mountains, stretching between 200 and 2060 m asl, a diachronic study of land use changes based on aerial photographs taken between 1945 and 1992 has shown considerable transformation, with an impressive decrease in arable lands, grasslands and very open shrublands and forests corresponding to an equally impressive increase in abandoned agricultural land and open and dense shrublands with forests (Table 8.1). Meanwhile, the active population of the mountain villages decreased by 32% and the number of sheep and goats fell by 18% between 1961 and 1991 (Chouvardas, 2001). Similar changes were also found in the Tymfi mountains of western Greece (Epirus) between 1945 and 1995, occurring especially between 1969 and 1995 (Zomeni *et al.*, 2008).

In the Lefka Ori of Crete, not only arable lands but also shrublands such as phrygana, garrigue and maquis declined in favour of coniferous forests, which expanded and became denser. Specifically, coniferous forests consisting of brutia pine (*Pinus brutia*) and cypress (*Cupressus sempervirens*) increased both in size and density between 1945 and 1989 (Papanastasis and Kazaklis, 1998). In the meantime, the human population decreased while sheep and goat numbers increased between 1961 and 1991 (Lyrintzis *et al.*, 1998). The Psiloritis mountain of Crete, in contrast, saw a decrease in dense shrubland and forests between 1961 and 1989 in favour of grasslands and especially the very open and open shrublands (Table 8.2). In the

Table 8.1   Land cover/use changes between 1945 and 1992 in the
Portaikos-Pertouli valley of the Pindos mountain range (central Greece)

Land cover/use type	Area (ha) 1945	1992	Change (%)
Artificial surfaces	84	147	+75.0
Arable lands	1877	1000	−46.7
Abandoned agricultural land	98	289	+194.9
Grasslands	1589	1084	−31.8
Shrublands:	975	872	+10.6
Very open (<40%)	418	119	−71.5
Open (40–70%)	302	400	+32.4
Dense (70–100%)	255	353	+38.4
Forests:	7594	8877	+16.9
Very open (<40%)	2304	1475	−36.0
Open (40–70%)	2697	2910	+7.9
Dense (70–100%)	2593	4492	+73.3
Barren land	232	180	−22.4
Total	12 449	12 449	0

Reproduced from Chouvardas (2001), courtesy of the author.

Table 8.2   Land cover/use changes between 1961 and 1989 in the
Psilorites mountain of Crete

Land cover/use type	Area (ha) 1961	1989	Change (%)
Artificial surfaces	245	245	0.0
Arable lands	13 575	11 634	−14.3
Grasslands	240	296	+23.3
Shrublands:	35 850	38 254	+6.7
Very open (<40%)	1590	2939	+84.8
Open (40–70%)	13 753	17 083	+24.2
Dense (70–100%)	20 507	18 232	−11.1
Forests:	5616	5097	−9.2
Very open (<40%)	3502	3212	−8.2
Open (40–70%)	2047	1840	−10.1
Dense (70–100%)	67	45	−32.8
Barren land	245	245	0.0
Total	55 771	55 771	0

Reproduced from Bankov (1998), courtesy of the author.

same period, although the human population remained unchanged, sheep and goat numbers increased almost four-fold. The latter increase explains why Psilorites was dominated by open and very open shrublands, which are degraded ecosystems composed of phryganic species that are unpalatable to livestock (Bankov, 1998). This represents an example of intensification of animal production in mountain areas leading to severe vegetation and land degradation. Thus livestock husbandry helped to maintain a relatively stable population in Psilorites but at the expense of the environment (Lyrintzis *et al.*, 1998).

8.4.7 Turkey

Most of the studies in Turkey report a small decline in forest cover suggesting that people are still in the mountains although some emigration has already started (Cakir *et al.*, 2007; Kadioğullari and Baskent, 2008; Günlü *et al.*, 2009). An opposite trend, namely a net 19.9% increase of the total forest area, was recorded in the Torul state forest between 1984 and 2005 due to the implementation of regeneration activities by the Forest Service and the emigration of the local population (Kadioğullari *et al.*, 2008). In the Mediterranean part of Turkey, traditional activities, especially farming, have started to be abandoned due to touristic development (Onur *et al.*, 2009). However, in the Mediterranean mountains, such as the Taurus, the traditional human practices, especially livestock husbandry, are still very active (Kaniewski *et al.*, 2007) (Figure 8.5).

Figure 8.5   A pastoral landscape in the Taurus Mountains of Turkey (photo by the author)

8.5 Discussion

8.5.1 Coupling land use with socioeconomic changes

From the description of the recent land use changes in the Mediterranean mountains it becomes clear that there are two main but opposite trends, which are shown in Figure 8.6. One is occurring in the southern Mediterranean, primarily in the Maghreb countries and secondarily in Turkey, where the trend is towards reduction of forest and rangeland cover in favour of arable farms. The other (opposite) trend is occurring in the northern Mediterranean, in southern Europe, where rangeland and especially forest cover is increasing at the expense of mountain farms (Figure 8.7). In the first case, the main driving force is population growth and the associated demand for arable and grazing land in the mountains for food production, while in the second case it is the rural depopulation and land abandonment. Between 1950 and 1998, the population of Morocco, Algeria and Tunisia increased by 300% while the population of Portugal, Spain, France and Italy grew by only 30% (Puigdefabregas

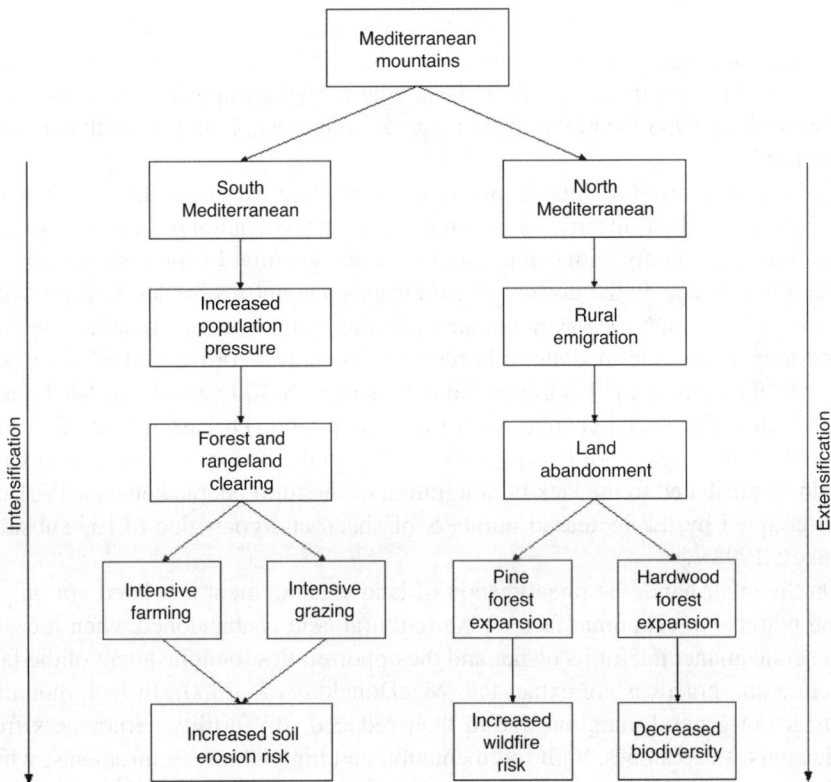

Figure 8.6 Generalized course of the land use changes in the mountains of the southern and northern Mediterranean countries

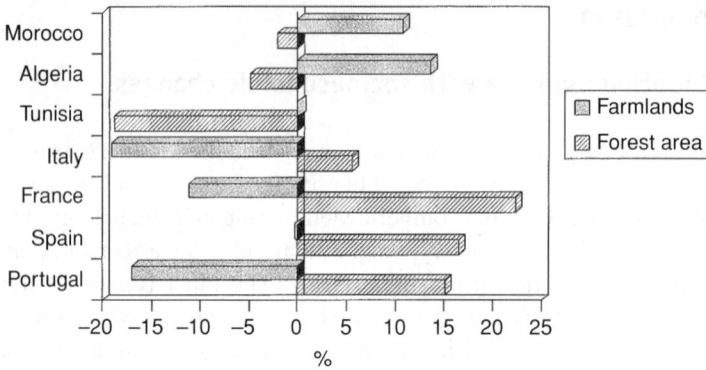

Figure 8.7 Transformation of land uses in western Mediterranean countries between 1965 and 1976. Reproduced from Mazzoleni *et al.* (2004a), with permission

and Mendizabal, 1998. For the whole Mediterranean, di Castri (1998) reports that the allocation of the total population of the countries bordering the Mediterranean Sea was 66.7% in the north and 33.3% in the south in 1950, but by 1990 it had become equal, and by 2025 it is projected to be 33.8% and 64.2% respectively, suggesting the high potential growth in the southern Mediterranean. This is due to the higher birth rate and the higher percentage of young people in the south compared to the north.

The pressure that this high population growth rate exerts on the mountains of the southern Mediterranean countries has led to intensification of land uses such as arable farming, mainly with cereals, and livestock grazing. In contrast, the depopulation of mountains in the northern Mediterranean countries has led to the extensification of land uses, such as mountain agriculture and livestock grazing, and land abandonment. An intermediate or borderline case was recorded in the Iblei mountains of Sicily where the landscape did not change for 100 years from 1897 due to the lack of major social changes over the same period (Di Pasquale *et al.*, 2004). Also, the intensification of livestock husbandry on the Psilorites mountain of Crete should be attributed to the lack of emigration of the rural people between 1961 and 1989 coupled by the increased numbers of sheep and goats due to EU subsidies (Bankov, 1998).

On the other hand, the phenomenon of land abandonment is related not only to social but also to economic factors. Agricultural land is abandoned when it ceases to generate an income for its owner and the opportunities for adjustment of the farm structure and practices are exhausted (MacDonald *et al.*, 2000). In fact, mountain lands are in general marginal due to their reduced soil fertility, remoteness from settlements, steep slopes, high fragmentation and high labour requirements, which limit their productivity and make their exploitation largely unprofitable. However, even if arable lands in the mountains are profitable, their cultivation is suspended if the owners are too old to continue cultivation, or they die, and no descendants are present or willing to take up farming. Another reason may be the multiple activities of the rural households, members of which are also engaged in non-farm activities such as tourism. The EU support of agriculture with subsidies in the southern European countries, although it provided a satisfactory income to the farmers, did not stop farm abandonment in the mountains, apparently because this did not lead to any structural improvements of agriculture (Kasimis and Papadopoulos, 2001).

8.5.2   Natural dynamics of land use changes

Although land use changes in the Mediterranean mountains are mainly caused by socioeconomic changes, environmental factors are also involved. They include climate, geology and relief. This is because a change in land use often produces a change in land cover that is directly affected by these factors. For example, abandoned land will be colonized by natural vegetation more rapidly under favourable rather than unfavourable climatic conditions. Also, climate conditions may reflect the types of vegetation that become established in the abandoned lands. According to Torta (2004), a prolonged shrubland phase occurs in the afforestation process after land abandonment in the northern Apennines, while in the western Alps this phase is absent or limited due to wetter conditions. He further states that climate, particularly the quantity and distribution of annual rainfall, plays a major role on a regional scale in determining the time and pace of afforestation after abandonment, while natural factors such as topography, geopedological conditions, slope and aspect play roles on a local scale. Such factors accounted for 13% of the total variance of the *Abies pinsapo* forest structure on the mountains of Spain and Morocco on opposite sides of the Strait of Gibraltar, while the contrasting land use (Spain vs Morocco) accounted for 23% of the total variance (Linares *et al.*, 2011).

The question that arises is whether plant or vegetation cover following the land use changes adheres to a standard theoretical model of change. After reviewing several studies on land use changes in Mediterranean countries, Mazzoleni *et al.* (2004b) concluded that the gradual reduction or disappearance of the original forest cover in the southern and eastern Mediterranean countries follows the model of successional regression. In the northern Mediterranean areas, on the contrary, land abandonment that triggers the increase of forest cover follows one of two models: (i) the auto-succession model in the areas where evergreen shrublands are colonized by pines, resulting in increased frequency and magnitude of wildfires; or (ii) the progression model in abandoned open spaces, forest plantations and sclerophyllous evergreen formations that leads to the original (pre-disturbance) forest cover composed of deciduous plant species (Figure 8.6).

Although there is a secondary succession of vegetation in the Mediterranean mountains following intensification of agriculture and livestock grazing or abandonment, this succession is not always linear. In the Psiloritis mountain of Crete, for example, where there is intensive grazing by sheep and goats, pine forests may be directly converted to phrygana (garrigues) without going through the maquis stage,

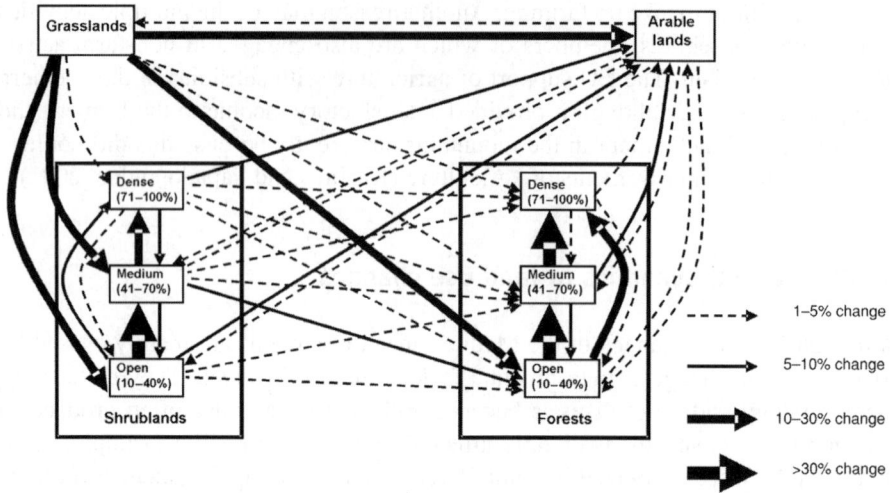

Figure 8.8 Multidirectional changes of land cover/use types between 1960 and 1993 in the Vertiskos mountain of northern Greece. Reproduced from Papanastasis and Chouvardas (2005), with permisson

or annual crops (cereals) may be directly colonized by pine trees without going first to grassland and then to shrubland (phrygana or maquis) (Bankov, 1998). Similar results were found in the southern French Alps, where the vegetation patterns of the European beach (*Fagus silvatica*) and silver fir (*Abies alba*) in the understorey of Scots pine (*Pinus silvestris*) after land abandonment were largely altered by grazing history, suggesting that agricultural land use and abandonment are both significant driving forces of vegetation dynamics (Chauchard *et al.*, 2007). The multiple pathways of land use changes are clearly shown in Figure 8.8 based on a study carried out in the Vertiskos mountain of northern Greece.

8.5.3 Environmental consequences of land use changes

Land use changes have several environmental impacts. One such impact is on landscapes. In the past, Mediterranean mountain landscapes, although shaped by human activities, had achieved an agro-silvo-pastoral equilibrium that helped serve both human needs and environmental conservation. This traditional equilibrium has been altered or disrupted in several Mediterranean mountain regions due to land use changes. In Alvão Natural Park of northern Portugal, for example, the landscape became more heterogeneous between 1947 and 1978 due to reforestation with conifers and deciduous trees, which led to an increase in the number of patches and diversity of the landscape. This diversity was further improved by wildfires and excessive livestock grazing leading to a new equilibrium with 60% of the area consisting of rangelands, 20% barren land and the remaining 20% forests and agricultural areas (Timóteo *et al.*, 2004). Increased landscape fragmentation was also recorded in northeastern Portugal between 1979 and 2002 (Pôcas *et al.*, 2011b). In contrast, the landscape became coarser and unifunctional in Serra de Monchique in southeast Portugal due to the total abandonment of rainfed agriculture and its replacement by eucalyptus afforestation (Krohmer and Deil, 2003). In the Spanish Pyrenees, although land use changes expanded the plant cover and contributed to a more natural landscape, they led towards simplification and homogenization of the landscape with decreased fragmentation resulting in an increased fire risk but decreased sediment transport to the lowlands (Lasanta-Martinez *et al.*, 2005). Homogenization of the landscape was also observed in the eastern Pyrenees after 1953, when mountain cropping and livestock grazing started to decrease in favour of forests (Roura-Pasqual *et al.*, 2005). A general homogenization of the landscape was also detected in Italy between 1954 and 2000 due to forest recovery (Geri *et al.*, 2010b); similar changes also took place in the White Mountains of Crete (Papanastasis and Kazaklis, 1998) and the Pindos mountains of central Greece (Chouvardas, 2001) due to land abandonment between 1945 and 1989. Interestingly, a less diverse and simpler landscape was also found on the Psilorites mountain of Crete, where land intensification (overgrazing) occurred between 1961 and 1989. This suggests that both extensification and intensification of land uses lead to homogeneous landscapes.

Land use changes also impact on biodiversity, both species and habitat diversity. Mediterranean mountains, whether in continental or island locations, are centres of plant diversity and endemism. According to Médail and Quezel (1999), the rates of plant species endemism exceed 20%. In addition, the same mountains are home to several endangered animal species, such as the Iberian lynx (*Lynx pardina*) in Spain and the Anatolian leopard (*Panthera pardus panthera*) in Turkey, as well as several endemic species such as the mouflon (*Ovis orientalis*) in Sardinia, Corsica and Cyprus, the Spanish ibex (*Capra pyrenaica*) (Regato and Salman, 2008) and the agrimi (*Capra aegagrus cretensis*) in the mountains of Crete. All these species are affected by land use changes. In general, land abandonment results in a decrease of plant species richness and habitats, particularly when the woodland canopy closes (MacDonald *et al.*, 2000) while the mammals and bird species associated with forests tend to increase (Falcucci *et al.*, 2007). However, plant species richness increases when livestock grazing and pastoral wildfires are extensified (Papanastasis *et al.*, 2002). On the other hand, carbon sequestration is increased with abandonment, especially when pastures are converted to woodlands (Padilla *et al.*, 2010).

Finally, land use changes have an impact on soils and natural hazards. Expansion of cultivated lands in forests and shrublands coupled with overgrazing of the remaining rangelands results in significant soil loss and sediment yields through erosion (Merzouk and Dhman, 1998; Puigdefabregas and Mendizabal, 1998). Land abandonment, on the other hand, and the resulting colonization of old arable fields by a dense shrub cover as well as afforestation can control both soil erosion and runoff (Garcia-Ruiz *et al.*, 1996; Garcia-Ruiz, 2010). However, if abandoned arable lands, particularly those on terraces, are intensively grazed by livestock then accelerated erosion may occur (Lasanta *et al.*, 2011). Moreover, the expansion of

forests and shrublands following land abandonment also has hydrological consequences, which are related to increased evapotranspiration and a reduction of run-
off (Puigdefabregas and Mendizabal, 1998). Conversely, a reduction in vegetation
cover through land use intensification results in the reduction of evapotranspiration
while run-off is increased thus increasing the risk of flooding and damage to engineering structures due to soil erosion and sedimentation (Mulligan *et al.*, 2004). Finally, the build-up of vegetation biomass after land abandonment increases the risk
of severe wildfires, which can also lead to soil erosion. In Spain, the area burned
annually increased significantly between 1960 and 1990, generating a debate about
the relation between wildfires and vegetation cover (Puigdefabregas and Mendizabal, 1998).

8.6   Conclusion

Although there are cases of massive deforestation in antiquity (e.g. the cypress
forests of Crete, cedar forests of Mount Lebanon), substantial land use changes in
the Mediterranean mountains started much later, in the medieval period, when wars,
famines and malaria in the lowlands forced people to move to the uplands. These
changes were intensified in the eighteenth and nineteenth centuries when populations increased and the Industrial Revolution greatly improved transport between
lowlands. The beginning of the twentieth century, however, marked the start of rural emigration from the mountains of northern Mediterranean countries that peaked
after the Second World War, resulting in land abandonment and the remarkable increase of forest cover seen in recent years. In southern Mediterranean countries, in
contrast, an opposite trend is recorded due to the substantial increase in the rural
population and consequent pressure on the mountains for arable and grazing land.
Recent land use changes have disrupted the traditional agro-silvo-pastoral equilibrium of the Mediterranean mountains. The increased forest cover in the northern
Mediterranean has resulted in a decline in species and especially of habitat diversity and an increase in natural hazards, especially wildfires, but increased carbon sequestration and decreased soil erosion and sediment transport to the lowlands. The
decrease of forest cover in the southern Mediterranean, on the other hand, has led
to severe soil erosion. A long-term monitoring system of all these land use changes
should be established and new equilibriums should be pursued in the Mediterranean
mountains so that their environment is conserved and the prosperity of mountain
people is better served.

8.7   Acknowledgements

Appreciation is expressed to the editor for his constructive comments.

References

References marked as bold are key references.

Ameztegni, A., Brotons, L. and Coll, L. (2010) Land use changes as major drivers of mountain pine (*Pinus uncinata* Ram.) expansion in the Pyrenees. *Global Ecology and Biogeography* 19:632–641.

Arianoutsou, M. (2001) Landscape changes in Mediterranean ecosystems of Greece: implications for fire and biodiversity issues. *Journal of Mediterranean Ecology* 2:165–178.

Bankov, N. (1998) Dynamics of land cover/use changes in relation to socio-economic conditions in the Psilorites mountain of Crete, Greece. MSc thesis, Mediterranean Agronomic Institute of Chania, Crete.

Bencherifa, A. (1983) Land use and equilibrium of mountain ecosystems in the High Atlas of western Morocco. *Mountain Research and Development* 3:273–279.

Bencherifa, A. and Johnson, D.L. (1991) Changing resource management strategies and their environmental impacts in the Middle Atlas Mountains of Morocco. *Mountain Research and Development* 11:153–194.

Biber, J.-P. (2010) Transhumance in France. *Pastoralism* 1:91–98.

Bonet, A., Bellot, J. and Pena, J. (2004) Landscape dynamics in a semiarid Mediterranean catchment (SE Spain). In Mazzoleni, S., di Pasquale, G., Mulligan, M., di Martino, P. and Rego, F. (eds), *Recent Dynamics of the Mediterranean Vegetation and Landscape*. Chichester: John Wiley & Sons, Ltd, pp. 47–56.

Butzer, K.W. (2003) The nature of Mediterranean Europe: an environmental history. *Annals of the Association of American Geographers* 93:494–498.

Cakir, G., Sivrikaya F., Terzioglou, S., Keles, S. and Baskent, E.Z. (2007) Monitoring thirty years of land cover change: secondary forest succession in the Artvin Forest Planning Unit at northeastern Turkey. *Scottish Geographical Journal* 123:209–225.

Chauchard, S., Carcaillet, C. and Guibal, F. (2007) Patterns of land use abandonment control tree recruitment and forest dynamics in Mediterranean mountains. *Ecosystems* 10:936–948.

Chouvardas, D. (2001) Analysis of temporal changes and structure of landscapes with the use of Geographic Information Systems (GIS). MSc thesis, Aristotle University of Thessaloniki, Greece [in Greek with English summary].

Cihlar, J. and Jansen, L.J.M. (2001) From land cover to land use: A methodology for efficient land use mapping over large areas. *Professional Geographer* 53:275–289.

Cohelho-Silva, J.L., Castro Rego, F., Castelbranco Silveira, S., Cardoso Goncalves, C.P. and Alberto Machado, C. (2004) Rural changes and landscape in Serra da Malcata, Central East of Portugal. In: Mazzoleni, S., di Pasquale, G., Mulligan, M., di Martino, P. and Rego, F. (eds), *Recent Dynamics of the Mediterranean Vegetation and Landscape*. Chichester: John Wiley & Sons, Ltd, pp. 189–200.

Conacher, A.J. and Sala, M. (1998) *Land Degradation in Mediterranean Environments of the World. Nature and Extent. Causes and Solutions.* New York: John Wiley & Sons, Ltd.

Davis, K.P. (1976) *Land Use.* New York: McGraw-Hill.

di Castri, F. (1998) Politics and environment in Mediterranean climate regions. In: Rundel, P.W., Montenegro, G. and Jaksic, F.M. (eds), *Landscape Disturbance and Biodiversity in Mediterranean-type Ecosystems*. Berlin: Springer, pp. 407–432.

Di Gregorio, A. and Jansen L.J.M. (1998) A new concept for a land classification system. *The Land* 2:55–65.

Di Pasquale, G., Gafri, G. and Migliozzi, A. (2004) Landscape dynamics in south-eastern Sicily in the last 150 years: the case of the Iblei mountains. In: Mazzoleni, S., di Pasquale, G., Mulligan, M., di Martino, P. and Rego, F. (eds), *Recent Dynamics of the Mediterranean Vegetation and Landscape*. Chichester: John Wiley & Sons, Ltd, pp. 73–80.

Douglas, T., Critchley, D. and Park, G. (1996) The deintensification of terraced agricultural land near Trevelez, Sierra Nevada, Spain. *Global Ecology and Biogeography* 5:258–270.

Etienne, M., Aronson, J. and Le Floc'h, E. (1998) Abandoned lands and land use conflicts in southern France. In: Rundel, P.W., Montenegro, G. and Jaksic, F.M. (eds), *Landscape Disturbance and Biodiversity in Mediterranean-type Ecosystems*. Berlin: Springer, pp. 125–140.

Falcucci, A., Maiorano, L. and Boitani, L. (2007) Changes in land use/land cover patterns in Italy and their implication for biodiversity conservation. *Landscape Ecology* 22:617–631.

FAO (1997) *Negotiating a Sustainable Future for Land: Structural and Institutional Guidelines for Land Resources Management in the 21st Century*. Rome: FAO/UNED.

Garcia-Ruiz, J.M. (2010) The effects of land use changes on soil erosion in Spain: A review. *Catena* 81:1–11.

Garcia-Ruiz, S.M., Lasanta, T., Ruiz Flano, P. *et al.* (1996) Land-use changes and sustainable development in mountain areas: a case study in the Spanish Pyrenees. *Landscape Ecology* 11:267–277.

Geri, F., Rocchini, D. and Chiarucci, A. (2010a) Landscape metrics and topographical determinants of large-scale dynamics in a Mediterranean landscape. *Landscape and Urban Planning* 95:46–53.

Geri, F., Amizi, V. and Rocchini, D. (2010b) Human activity impact on the heterogeneity of Mediterranean landscape. *Applied Geography* 30:370–379.

Gracia, M., Meghelli, N., Comas, L. and Retana, J. (2011) Land-cover changes in and around a National Park in a mountain landscape in the Pyrenees. *Regional Environmental Change* 11:349–358.

Grove, A.T. and Rackham, O. (2001) *The Nature of Mediterranean Europe. An Ecological History*. New Haven, CT: Yale University Press.

Günlü, A., Kadiogullari, A.I., Keles, S. and Baskent, E.Z. (2009) Spatiotemporal changes of landscape pattern in response to deforestation in northeastern Turkey: a case study in Rize. *Environmental Monitoring and Assesment* 48:127–137.

Hughes, J.D. (1983) How the ancients viewed deforestation. *Journal of Field Archaeology* 10:435–445.

Ispikoudis, I., Sioliou, M.K. and Papanastasis, V.P. (2004) Transhumance in Greece: past, present, and future prospects. In: Bunce, R.G.H., Perez-Soba, M., Jongman, R.H.G., Gomez Sal, A., Herzog, F. and Austad, I. (eds), *Transhumance and Biodiversity in European Mountains*. IALE Publication Series no. 1, pp. 211–229.

Kadioğullari, A.I. and Baskent, E.Z. (2008) Spatial and temporal dynamics of land use pattern in eastern Turkey: a case study in Gümüshane. *Environmental Monitoring and Assessment* 138:289–303.

Kadioğullari, A.I., Keles, S., Baskent, E.Z. and Günlü, A. (2008) Spatiotemporal changes in response to afforestation in northeastern Turkey: a case study of Torul. *Scottish Geographical Journal* 214:259–273.

Kaniewski, D., De Laet, V., Paulissen, E. and Waelkens, M. (2007) Long-term effects of human impact on mountainous ecosystems, western Taurus mountains, Turkey. *Journal of Biogeography* 34:1975–1997.

Kasimis, C. and Papadopoulos, A.G. (2001) The de-agriculturisation of the Greek countryside: the changing characteristics of an ongoing socio-economic transformation. In: Cranberg, L., Kovach, I. and Tovey, H. (eds), *Europe' s Green Ring*. Aldershot: Ashgate, pp. 197–219.

Körner, C. and Ohsawa, M. (coords. lead authors) (2005) Mountain systems. In: Hassan, R., Scholes, R. and Ash, N. (eds), *Millennium Ecosystem Assessment*. Island Press, chapter 24.

Krohmer, J. and Deil, U. (2003) Dynamic and conservative landscapes? Present vegetation cover and land-use changes in the Serra de Monchique (Portugal). *Phytocoenologia* 33:767–799.

Lasanta, T., Laguta, M. and Vicente Serrano, S.M. (2007) Do tourism-based ski resorts contribute to the homogeneous development of the Mediterranean mountains. A case study in the Central Spanish Pyrenees. *Tourism Management* 28:1328–1339.

Lasanta, T., Armaez, J., Oserin, M. and Ortigosa, M. (2011) Marginal lands and erosion in terraced fields in the Mediterranean mountains. *Mountain Research and Development* 21: 69–76.

Lasanta-Martinez, T., Vicente-Serrano, S.M. and Cuadrat-Prats, J.M. (2005) Mountain Mediterranean landscape evaluation caused by the abandonment of traditional primary activities: a study of the Spanish Central Pyrenees. *Applied Geography* 25:47–65.

Liacos, L.G. (1973) Present studies and history of burning in Greece. In: *Proceedings of the 13th Annual Tall Timbers Fire Ecology Conference*. Tall Timbers Research Station, Tallahassee, Florida, pp. 65–95.

Linares, J.S., Carreira, J.A. and Ochoa, V. (2011) Human impacts drive forest structure and diversity. Insights from Mediterranean mountain forests dominated by *Abies pinsapo* (Boiss). *European Journal of Forestry Research* 130:533–542.

Lyrintzis, G. and Papanastasis, V. (1995) Human activities and their impact on land degradation – Psilorites mountain in Crete: An historical perspective. *Land Degradation and Rehabilitation* 6:79–93.

Lyrintzis, V.P., Papanastasis, V.P. and Ispikoudis, I. (1998) Role of livestock husbandry in social and landscape changes in White Mountains and Psilorites of Crete. In: Papanastasis, V.P. and Peter, D. (eds), *Ecological Basis of Livestock Grazing in Mediterranean Ecosystems*. Luxembourg: European Commission, EUR 18308, pp. 322–327.

MacDonald, D., Crobtree, J.R., Wiesinger, G. *et al.* (2000) Agricultural abandonment in mountain areas of Europe: environmental consequences and policy response. *Journal of Environmental Management* 59:47–69.

Mazzoleni, S., di Martino, P., Strumia, S., Buonanno, M. and Bellelli, M. (2004a) Recent changes of coastal and sub-mountain vegetation landscape in Campania and Molise regions in Southern Italy. In: Mazzoleni, S., di Pasquale, G., Mulligan, M., di Martino, P. and Rego, F. (eds), *Recent Dynamics of the Mediterranean Vegetation and Landscape*. Chichester: John Wiley & Sons, Ltd, pp. 145–155.

Mazzoleni, S., di Pasquale, G. and Mulligan, M. (2004b) Conclusion: reversing the consensus on Mediterranean desertification. In: Mazzoleni, S., di Pasquale, G., Mulligan, M., di Martino, P. and Rego, F. (eds), *Recent Dynamics of the Mediterranean Vegetation and Landscape*. Chichester: John Wiley & Sons, Ltd, pp. 281–285.

McNeill, J.R. (1992) *The Mountains of the Mediterranean World: An Environmental History*. New York: Cambridge University Press.

Médail, F. and Quezel, P. (1999) Biodiversity hotspots in the Mediterranean basin: setting global conservation priorities. *Conservation Biology* 13:1510–1513.

Meiggs, R. (1982) *Trees and Timber in the Ancient Mediterranean World*. Oxford: Clarendon Press.

Merzouk, A. and Dhman, H. (1998) Shifting land use and its implication on sediment yield in the Rif mountains (Morocco). *Advances in Geoecology* 31:333–340.

Moreno, J.M., Vázquez, A. and Vélez, R. (1998) Recent history of forest fires in Spain. In: Moreno, J.M. (ed.), *Large Forest Fires*. Leiden: Backhuys Publishers, pp. 159–185.

Motta, R., Morales, M. and Nola, P. (2006) Human land use, forest dynamics and tree growth in the western Italian Alps. *Annals of Forest Science* 63:739–749.

Mulligan, M., Burke, S.M. and Ramos, M.C. (2004) Climate change, land use change and the "desertification" of Mediterranean Europe. In: Mazzoleni, S., di Pasquale, G., Mulligan, M., di Martino, P. and Rego, F. (eds), *Recent Dynamics of the Mediterranean Vegetation and Landscape*. Chichester: John Wiley & Sons, Ltd, pp. 259–279.

Naveh, Z. (1975) The evolutionary significance of fire in the Mediterranean region. *Vegetatio* 29:199–208.

Naveh, Z. and Lieberman, A.S. (1994) *Landscape Ecology. Theory and Applications*, 2nd edn. New York: Springer-Verlag.

Onur, I., Maktan, D., Sari, M. and Sommez, N.K. (2009) Change detection of land cover and land using remote sensing and GIS: a case study in Kemer, Turkey. *International Journal of Remote Sensing* 30:1749–1757.

Padilla, F.M., Vidal, B., Sánchez, J. and Pugnaire, F.I. (2010) Land use changes and carbon sequestration through the twentieth century in a Mediterranean mountain ecosystem: Implications for land management. *Journal of Environmental Management* 91:2688–2695.

Papanastasis, V.P. (1998) Livestock grazing in Mediterranean ecosystems: an historical and policy perspective, In: Papanastasis, V.P. and Peter, D. (eds), *Ecological Basis of Livestock Grazing in Mediterranean Ecosystems*. Luxembourg: European Commission, EUR 18308, pp. 5–9.

Papanastasis, V.P. (2004) Traditional vs contemporary management of Mediterranean vegetation: the case of the island of Crete. *Journal of Biological Research–Thessaloniki* 1:39–46.

Papanastasis, V.P. (2007) Land abandonment and old field dynamics in Greece. In: Cramer, V.A. and Hobbs, R.J. (eds), *Old Fields: Dynamics and Restoration of Abandoned Farmland*. London: Island Press, pp. 225–246.

Papanastasis, V.P. and Chouvardas, D. (2005) Application of the state-and-transition approach to conservation management of a grazed Mediterranean landscape in Greece. *Israel Journal of Plant Sciences* 53:191–202.

Papanastasis, V.P. and Kazaklis, A. (1998) Landuse changes and conflicts in the Mediterranean-type ecosystems of western Crete. In: Rundel, P.W., Montenegro, G. and Jaksic, F.M. (eds), *Landscape Disturbance and Biodiversity in Mediterranean-type Ecosystems*. Berlin: Springer, pp. 141–157.

Papanastasis, V.P., Kyriakakis, S. and Kazakis, G. (2002) Plant diversity in relation to overgrazing and burning in mountain Mediterranean ecosystems. *Journal of. Mediterranean Ecology* 2:53–63.

Papanastasis, V.P., Arianantsou, M. and Papanastasis K. (2010) Environmental conservations in classical Greece. *Journal of Biological Research–Thessaloniki* 14:123–135.

Pôcas, I., Cunha, M., Marcal, A.R.S. and Pereira, L.S. (2011a) An evaluation of changes in a mountainous rural landscape of northeast Portugal using remotely sensed data. *Landscape and Urban Planning* 101:253–261.

Pôcas, I., Cunha, M. and Pereira, L.S. (2011b) Remote sensing based indicators of changes in a mountain rural landscape of northeast Portugal. *Applied Geography* 31:871–880.

Poyatos, R., Latron, S. and Llorens, P. (2003) Land use and land cover change after agricultural abandonment. The case of a Mediterranean mountain area (Catalan Pre-Pyrenees). *Mountain Research and Development* 23:362–368.

Puigdefabregas, J. and Mendizabal, T. (1998) Perspectives on desertification: western Mediterranean. *Journal of Arid Environments* 39:204–224.

Rackham, O. and Moody, J.A. (1992) Terraces. In: Wells, B. (ed.), *Agriculture in Ancient Greece*. Proceedings of the Seventh International Symposium at the Swedish Institute at Athens. Scrifter Utgivna av Svenska Institute i Athen. Stockholm: Paul Äströms Förlag, pp. 123–133.

Regato, P. and Salman, R. (2008) *Mediterranean Mountains in a Changing World. Guidelines for Developing Action Plans*. Malaya, Spain: JUCN Centre for Mediterranean Cooperation, 88 pp.

Regato-Pajares, P., Jimenez-Caballero, S., Castejoni, M. and Elena-Rossello, R. (2004) Recent landscape evolution in dehesa woodlands of western Spain. In: Mazzoleni, S., di Pasquale, G., Mulligan, M., di Martino, P. and Rego, F. (eds), *Recent Dynamics of the Mediterranean Vegetation and Landscape*. Chichester: John Wiley & Sons, Ltd, pp. 57–72.

Rejdali, M. (2004) Forest cover changes in the Maghreb countries with special reference to Morocco. In: Mazzoleni, S., di Pasquale, G., Mulligan, M., di Martino, P. and Rego, F. (eds), *Recent Dynamics of the Mediterranean Vegetation and Landscape*. Chichester: John Wiley & Sons, Ltd, pp. 23–31.

Roura-Pasqual, N., Pons, P., Ettiene, M. and Lambert, B. (2005) Transformation of a rural landscape in the eastern Pyrenees between 1953 and 2000. *Mountain Research and Development* 5:252–261.

Ruiz, M. and Ruiz, J.P. (1986) Ecological history of transhumance in Spain. *Biological Conservation* 37:73–86.

Ruiz, M. and Valero, A. (1990) Transhumance with cows as a rational land use option in the Gredos Mountains (Central Spain). *Human Ecology* 18:187–201.

Silva, J.S., Vaz, P., Moreira, F., Catry, F. and Rego, F.C. (2011) Wildfires as a major driver of landscape dynamics in three fire-prone areas of Portugal. *Landscape and Urban Planning* 101:349–358.

Susmel, P., Fabro, C., and Filacorda, S. 2004. Transhumance in the Italian Alps and Apennines. In: Bunce, R.G.H., Perez-Soba, M., Jongman, R.H.G., Gomez Sal, A., Herzog, F. and Austad, I. (eds), *Transhumance and Biodiversity in European Mountains*. IALE Publication Series no. 1, pp. 231–232.

Taillefumier, F. and Piégan, H. (2003) Contemporary land use changes in pre-alpine Mediterranean mountains: a multivariate GIS-based approach applied to two municipalities in the southern French Alps. *Catena* 51:267–296.

Tatoni, T., Médail, F., Roche, P. and Barbero, M. (2004) The impact of changes in land use on ecological patterns in Provence (Mediterranean France). In: Mazzoleni, S., di Pasquale, G., Mulligan, M., di Martino, P. and Rego, F. (eds), *Recent Dynamics of the Mediterranean Vegetation and Landscape*. Chichester: John Wiley & Sons, Ltd, pp. 107–120.

Thirwood, J.V. (1981) *Man and the Mediterranean Forest. A History of Resource Depletion*. New York: Academic Press.

Timóteo, I., Bento, J., Castro Rego, F. and Fernandes, A. (2004) Changes in landscape structure of the Natural Park of Alvao (Portugal). In: Mazzoleni, S., di Pasquale, G., Mulligan, M., di Martino, P. and Rego, F. (eds), *Recent Dynamics of the Mediterranean Vegetation and Landscape*. Chichester: John Wiley & Sons, Ltd, pp. 211–216.

Torta, G. (2004) Consequences of rural abandonment in a Northern Apennines landscape (Tuscany, Italy). In: Mazzoleni, S., di Pasquale, G., Mulligan, M., di Martino, P. and Rego, F. (eds), *Recent Dynamics of the Mediterranean Vegetation and Landscape*. Chichester: John Wiley & Sons, Ltd, pp. 157–165.

Tsoumis, G. (1986) The depletion of forests in the Mediterranean region. A historical review from ancient times to the present. *Scientific Annals of the Department of Forestry and Natural Environment. Aristotelian University of Thessaloniki* 28:282–300.

Vakalopoulos, A. (1964) *History of Modern Greece: Ottoman Period (1453–1669)* [in Greek]. Thassaloniki, Greece: Ant. Stamoulis.

van Andel, T.H., Zangger, E. and Demitrack, A. (1990) Land use and soil erosion in prehistoric and historical Greece. *Journal of Field Archaeology* 17:379–396.

Zomeni, M., Tzanopoulos, J. and Pantis, J.D. (2008) Historical analysis of landscape change using remote sensing techniques: an explanatory tool for agricultural transformation in Greek rural areas. *Landscape and Urban Planning* 86:38–46.

9

Climate change and its impact

D. Nogués-Bravo, J.I. López-Moreno and S.M. Vicente-Serrano

9.1 Introduction

We are currently experiencing a period of global climatic changes that are affecting the biosphere and the ecosystem services that sustain our societies. The relationships and feedbacks between the different abiotic and biotic elements that compound the Earth's global ecosystem, such as climate, oceans, glaciers and rivers, biodiversity and human societies, are usually complex and non-linear. In any case, climate and climate changes through time are clearly the engine that modifies our environment – its fingerprints present all over the world (Parmesan and Yohe, 2003; Root *et al.*, 2003) – and mountains are no exception to this rule. Mountains are amongst the most fragile environments in the world (Diaz *et al.*, 2003). They are a repository of biodiversity, water and other ecosystem services (Korner, 2004; Viviroli and Weingartner, 2004; Woodwell, 2004), and their influence exceeds their geographical limits and extends to the surrounding lowlands. Approximately 26% of the world's human population inhabit mountainous regions (Meybeck *et al.*, 2001). Mountains constitute centres of endemism for biodiversity, harbouring endangered species and habitats. Mountains also provide services with tangible economic value, such as power supplies, tourism, crop and livestock production, and perhaps the most important among them, water, since mountains are the source of approximately 70% of fresh water in the world (Viviroli *et al.*, 2007).

Climate change could greatly affect these ecosystem services in the coming decades. However, climatic changes and their influence on the environment are not confined to recent times. On the contrary, the Earth's global climatic system has been changing continuously for billions of years and the magnitude and impacts of such change on mountain ranges have been explored and documented intensively. This short digression to past climatic changes allows us to place current climate change and its impacts in a proper context, avoiding on the one hand dramatic interpretations but, on the other hand, recognizing the ability of climatic changes to radically transform physical, biological and social realms. Projections of future

Mediterranean Mountain Environments, First Edition. Edited by Ioannis N. Vogiatzakis.
© 2012 John Wiley & Sons, Ltd. Published 2012 by John Wiley & Sons, Ltd.

climatic changes in Mediterranean mountains (Nogués-Bravo *et al.*, 2008a) predict significant warming of between 1.6°C and 8.3°C for 2085, and even more importantly, climate models also project a reduction of precipitation, mainly during spring (17% under A1FI and 4.8% under B1 for 2085; see Figure 9.1 for an explanation on emission scenarios). This is of utmost importance as Mediterranean mountains are the main sources of water supply to their usually dry and highly populated surrounding lowlands. In this context, this chapter intends to assess the recent and future trends of climate in Mediterranean mountains, to summarize recent impacts and to couple the future implications of predicted climate change for human and physical features. Specifically, the first part of this chapter is devoted to summarizing recent trends of key climatic variables such as temperature of rainfall, and to compare them against future projections. Then, we summarize impacts of climate change on the cryosphere and the hydrosphere, and finally we discuss climate change impacts on biodiversity.

9.2   Climate change in Mediterranean mountains

Mediterranean mountains are expected to become warmer and to suffer a reduction of rainfall during the twenty-first century (Nogués-Bravo *et al.*, 2008a). The magnitude of this future climate change varies in relation to the economic, political and technological future evolution of our societies; in other words, it varies in relation to the different narrative storylines used to develop IPCC emission scenarios(Figure 9.1). A future world with very rapid economic growth and intensive fossil fuel use (A1FI could be considered a pessimistic scenario for the future in terms of climate change) yields the warmest climate change scenario (3.18°C in 2055 and 5.28°C in 2085). Even when we use the most optimistic scenario, the expected warming rates (2.28°C and 2.88°C under B1 in 2055) will become markedly more intense than the observed 0.76°C warming for the Mediterranean Basin reported by Giorgi (2002) for the twentieth century. Similarly, predictions for the twenty-first century are much larger than the 0.98°C warming observed, for example, in the Pyrenees during the twentieth century (Bucher and Dessens, 1991). Also, climate models project reductions of annual and spring precipitation in the Mediterranean mountains (e.g. 11.3% and 17% during spring under a pessimistic scenario for 2055 and 2085, respectively). Thus, Mediterranean mountain ecosystems (structure, functions and services) will likely be subject to intensive transformation in the coming decades, even assuming the most conservative estimates.

These projections are based on high-resolution Atmospheric-Ocean-coupled General Circulation Models, AOGCMs. Downscaled climate models for specific regions of the world, such as Regional Climate Models (RCMs), show similar trends, and therefore confirm the ability of high-resolution AOGCMs to forecast climate change in topographically complex areas. Specifically for the Pyrenees,

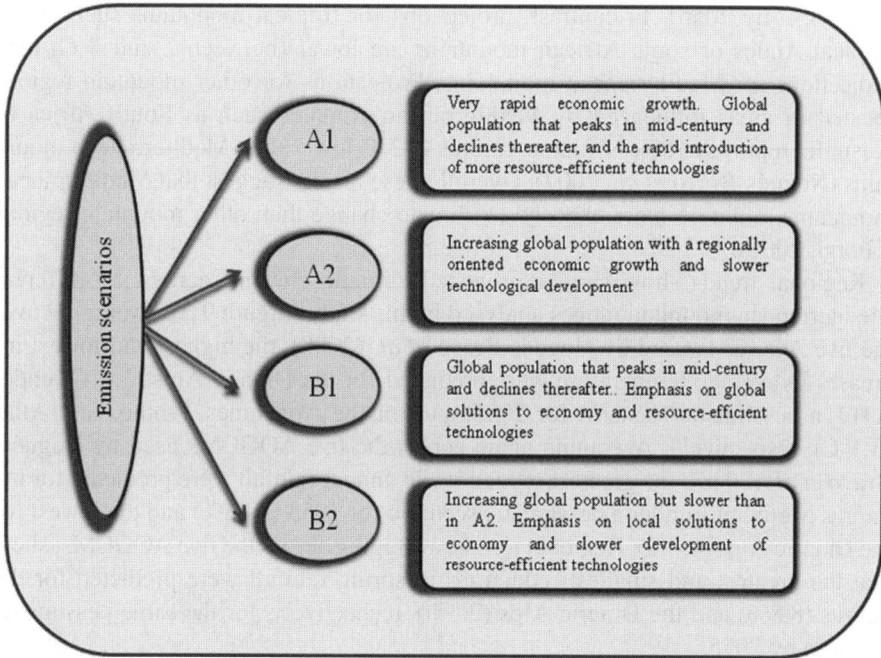

Figure 9.1 Four families of emission scenarios for greenhouse gases proposed by the Intergovernmental Panel on Climate Change (IPCC, 2007) corresponding to four different possible future 'worlds'. Assumptions about future socioeconomic and technological developments are made for each scenario

both AOGCMs and RCMs project similar changes for average annual temperature, annual precipitation, and spring precipitation for 2085. Projected warming using the AOGCM HadCM3 is 4.28°C and 3.28°C under a pessimistic and an optimistic scenario, respectively. The average warming projected by the RCMs will vary between 3.58°C and 2.58°C. Similarly, the reduction of annual precipitation projected by the AOGCM varies between 18.9% and 10% for B2, and the average reduction projected by the RCMs is 13% and 9%, respectively. Finally, for the spring precipitation, the reduction projected by the AOGCM varies between 21% and 16%, and the average reduction projected by the RCMs is 16% and 8%, respectively.

A specific trend in Mediterranean mountains is that they will suffer a reduction in annual rainfall and in spring rainfall, as reported above. In contrast, the predictions for non-Mediterranean European mountain regions project a slight increase of roughly 10% in both parameters by 2085. Also, Mediterranean mountains, when compared to other mountain ranges, will suffer a similar magnitude of warming (Nogués-Bravo *et al.*, 2007) to those mountains located in temperate regions of the world, such as the Rocky Mountains or the Himalayas (between

3 and 4°C by 2055). In contrast, projections for tropical mountains such as the tropical Andes or some African mountains are lower (between 2 and 3°C) than projections for Mediterranean mountains. Projections for other mountain regions located in areas influenced by Mediterranean climates such as South Africa or Australia report increases of temperatures 1–2°C lower than Mediterranean mountains (Nogués-Bravo *et al.*, 2007). Overall, these results suggest that Mediterranean mountains might be more exposed to climate change than other mountain regions (Giorgi, 2006).

Regional trends show similar projected climatic changes across the different Mediterranean mountain ranges analysed in this study (Figure 9.2). Averaging over the five AOGCMs used by Nogués-Bravo *et al.* (2008), the highest and lowest increases in average temperature were predicted for the Dinaric Alps (3.4°C) under A1FI, a pessimistic scenario for 2055, and for the Apennines, Pinthos and Atlas (3.1°C), respectively. Averaging again across the five AOGCMs used by Nogués-Bravo *et al.* (2008), the greatest reductions in annual rainfall were predicted for the Taurus Mountains under a pessimistic scenario for 2055 (14.8%) and the lowest for the Dinaric Alps (8.8%). Regional trends, averaging across the five AOGCMs, show that the greatest and smallest reductions in spring rainfall were predicted for the Taurus (8.8%) and the Dinaric Alps (2.3%), respectively, for the same pessimistic scenario by 2055.

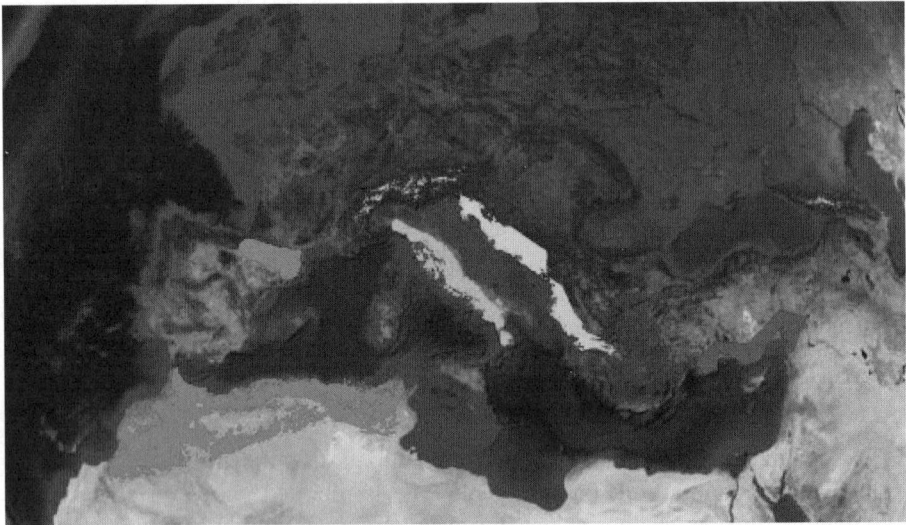

Figure 9.2 Mountainous areas analyzed in this chapter. Mountain areas have been defined following the UNEP-WCMC classification. Analyses of climatic and Normalized Difference Vegetation Index (NDVI) trends are just within the strict limits of the mountain ranges in this figure. Pyrenees (pink), Apennines (orange), Dinaric Alps (green), Pindos (blue), Taurus (purple) and Atlas (red). (*A full colour version of this figure appears in the colour plate section*)

Climate change predictions do not come without assumptions and uncertainties, and therefore their projections should be interpreted carefully. For example, an average warming of 3°C may well mask the real results used to calculate this average, or in simple terms, a model realization could predict a 1°C increase in temperature and another realization a 5°C increase, with 3°C being the average value. This huge difference among different projections and their influence on predicted impact needs more studies to explain the degree of uncertainty in climate projections. For example, predicted changes of precipitation in Mediterranean mountains differ noticeably among AOGCMs and emission scenarios (Nogués-Bravo *et al.*, 2008a). For 2055, the greatest reduction of average annual precipitation for Mediterranean mountains was 12.5%, whereas other models projected an increase of 0.9%. A similar pattern was recorded for 2085, although the difference among projections was greater for 2085 (23% and 1.4%, respectively).

In the paragraphs above we have summarized the projected climatic trends for the twenty-first century, which show mainly an increase in temperature and a decrease in rainfall. This trend was also significant during the twentieth century. To assess this climatic trend, we have analyzed the moisture trends in the six mountain ranges using a drought index: the Standardized Precipitation Evapotranspiration Index (SPEI) (Vicente-Serrano *et al.*, 2010). This index combines the effect of precipitation inputs and temperature in the moisture conditions. Using a climatic dataset (P.D. Jones *et al.*, in preparation; http://badc.nerc.ac.uk/data/cru/) from 1901 to 2006, we obtained annual (January to December) SPEI values for each one of the 0.5° pixels corresponding to each of the six mountain ranges analyzed here. Trends in moisture for each mountain range were calculated using the non-parametric Mann–Kendall test ($p < 0.05$). Table 9.1 shows the percentage of surface in each mountain range with positive (increase of moisture conditions) and negative trends (decrease of moisture conditions), and a clear domain of negative and significant trends in most of the mountain ranges. It is remarkable that those mountain ranges located in the westernmost areas (Atlas, Pyrenees and Apennines) show the highest percentage of the surface area with negative and significant trends. Thus, the Atlas shows the clearest decrease in moisture conditions among the six analyzed mountain ranges, with 95.6% of the surface affected by negative trends (see also Figure 9.3 showing the evolution of the SPEI in each of the six mountain ranges).

Table 9.1 Percentage of the surface area with negative and positive trends in the Standardized Precipitation Evapotranspiration Index (SPEI) values for each mountain range

Trend	Pyrenees	Apennines	Dinaric Alps	Pindos	Taurus	Atlas
Negative (sig.)	85.7	67.7	8.3	35.7	51.7	95.6
Negative (non-sig.)	14.3	32.3	75.0	64.3	48.3	4.4
Positive (non-sig.)			16.67			

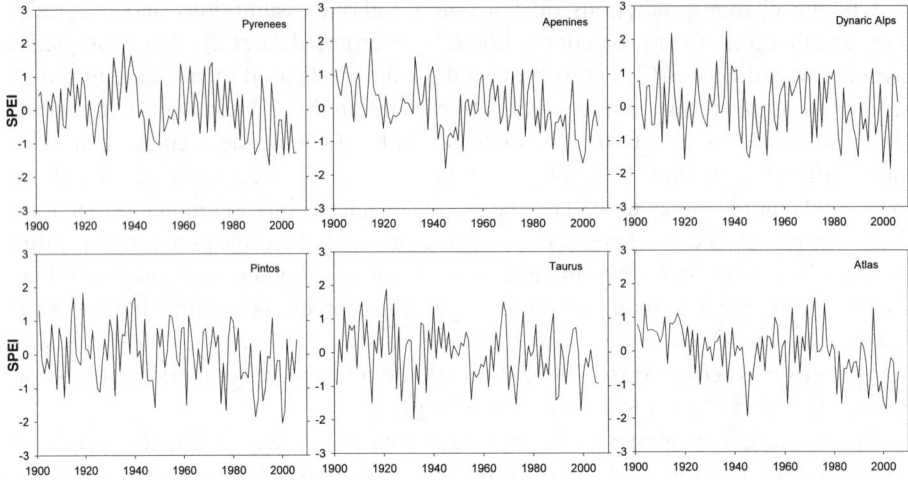

Figure 9.3   The evolution of the SPEI (Standardized Precipitation Evapotranspiration Index) in each of the six mountain ranges, obtained from the average climate series in each mountainous region. A falling trend in SPEI is observed in all regions, and this is most apparent and sustained in the Pyrenees, Apennines and the Atlas

In summary, Mediterranean mountains will become significantly warmer and drier during this century, with the western Mediterranean mountains suffering the greatest decreases in moisture conditions.

9.3   Cryosphere changes in the Mediterranean mountains

Warming temperatures and a generalized tendency to lower precipitation has led to consistent changes in snowfall and the snowpack extent in mountainous Mediterranean areas, although this trend is not specific to Mediterranean mountains. For example, Marty (2008) found a marked decrease in snow accumulation in the Alps during the last two decades, separated from previous accumulation values by an abrupt change that occurred around 1980. A decrease since the 1940s in the duration of snow cover has been reported for the Iberian Peninsula (Sanz-Elorza *et al.*, 2003), and in the Spanish Pyrenees. López-Moreno (2005) showed a clear trend of decrease from 1950 to 2000, associated with a fall in precipitation and atmospheric circulation changes, during February and March (López-Moreno and Vicente-Serrano, 2007). Changes in snowpack have been noticeably greater at low altitudes as a consequence of the rising of the 0°C isotherm (Beniston *et al.*, 2003).

Warmer temperatures and reduced snow accumulation during winter and spring have caused a marked retreat of glaciers in the Mediterranean region. Smaller glaciers in the Apennines and Pyrenees have retreated dramatically and they are currently close to complete disappearance (D'Orefice *et al.*, 2000; López-Moreno *et al.*, 2006). Other mountain ranges next to the Mediterranean Basin, such as the Alps, are among those exhibiting the greatest glacier retreats worldwide (Oerlemans, 2005). Thus, a dramatic degradation in Mediterranean glaciers is very likely under the warming magnitudes projected by climate models.

9.4   Hydrosphere changes in the Mediterranean mountains

Mountains provide various ecosystem services (Nogués-Bravo *et al.*, 2008a), both within the mountain ranges and also to neighbouring lowlands, and among them the most important in Mediterranean mountains is water. Therefore, the hydrosphere changes in mountain ranges drastically affect water supply across the whole Mediterranean basin. Specifically, a significant relationship has typically been observed across all temporal scales (regardless of the basin area studied or geomorphological and geological characteristics) between the temporal evolution of precipitation, at annual or longer scales, and river discharge.

The role of temperature in the evolution of water resources is generally more difficult to assess than that of precipitation. However, some studies have confirmed that warming has a role in explaining recent trends in water resources in the Mediterranean Basin. Thus, Pandžić *et al.* (2009) attributed the decrease since 1950 in discharge from the Sava River (Croatia) to a negative trend in soil moisture availability, which is closely related to increased evapotranspiration. A similar conclusion was found for the headwaters of the Duero River in the Iberian Peninsula (Ceballos *et al.*, 2008).

As expected from the changes in precipitation and temperature, a generalized and consistent decrease in river discharge has been reported in all studies concerning this aspect of the Mediterranean Basin (López-Moreno *et al.*, 2008).

The occurrence of severe droughts has affected the duration and severity of low flow periods, which have significant ecological, economic and social impact. For instance, Zanchettin *et al.* (2008) analyzed the evolution of stream flow in the Po River, northern Italy, since the beginning of the nineteenth century; no changes in average annual discharge during the twentieth century were found. Nevertheless, the cited authors found an increase in peak flow discharge in recent decades, and a higher frequency of drought periods since 1940. Snow accumulation and snowmelt play a determining role in the seasonal distribution of river discharge, especially in mountain basins in the Alps, the Carpathians, the Balkans, and the Pyrenees. As a consequence, higher discharge is usually expected in spring, when snowmelt and rainfall combine to give high flows between April and June.

Global warming affects snow accumulation and snowmelt in two ways: (i) reducing snow accumulation in the headwaters and giving more rain events in winter; and (ii) causing earlier thawing in the season. Both responses cause major changes in river regimes (Beniston *et al.*, 2003). For instance, López-Moreno and García-Ruiz (2004) reported a large effect of decreasing snowpack on changes in Pyrenean river regimes, particularly a clear decrease in the spring high flows and the occurrence of the high flows at least 1 month earlier than heretofore. Another important consequence is that water reserves in headwaters are exhausted earlier, with declining summer low flows forcing reservoir managers to reduce the discharge from dams (López-Moreno *et al.*, 2008).

For the highest headwaters in a mountain range next to the Mediterranean Basin, such as the Alps, an increase in temperature can result in rises for discharge related to glacier melting, particularly in summer, as was observed in the southern French Alps by Renard *et al.* (2008). These processes affect only small sectors of the river basins, but jeopardize the sustainability of high mountain water resources.

Climate change projected for the region will severely impact river discharge and water resource availability. Even if precipitation remains stable, an increase of temperature may cause a noticeable decline in stream flow as a consequence of enhanced evapotranspiration (Nash and Gleick, 1993). Despite the uncertainty associated with projections for precipitation, there is robust evidence for predicting a decrease in this variable affecting the majority of the Mediterranean Basin. Less precipitation will reinforce the likely effects of warming temperatures in reducing water resources (Manabe *et al.*, 2004).

Some studies based on hydrological models and the outputs from different scenarios obtained from climate change models point towards a diminution of surface run-off and groundwater recharge (Younger *et al.*, 2002; Nunes *et al.*, 2008). Fujihara *et al.* (2008) have shown that the drought return period for the Sayehan River basin (Turkey) will change from 5.3 years under present conditions to 2.0 years under future conditions; that is, critical hydrological drought events will occur more frequently as a consequence of climate change. Overall, climate change scenarios suggest a marked increase of water resource stress in the Mediterranean and the Middle East (Arnell, 2004).

9.5   Biosphere changes in the Mediterranean mountains

The Mediterranean Basin is one of the global biodiversity hotspots (Myers, 2003) and it harbours, for example, more than 22 500 endemic vascular plant species: more than four times the number found in all the rest of Europe. One of the main reasons behind the high biological diversity in the Mediterranean Basin is the high climate stability across different timescales. The Mediterranean Basin provided refugia during the Pleistocene cycles, preventing species extinctions and likely supporting speciation processes. Moreover, the Mediterranean mountains played a key role in this, providing 36% (18 out of 52) of the recently proposed glacial refugia within the Mediterranean Basin (Médail and Diadema, 2009), among which are the Pyrenees, South Apennines, Taurus and Atlas.

Recent studies have suggested that changes in temperature and precipitation would lead to a shift towards vegetation types currently found under drier and warmer conditions in Mediterranean mountains (Gritti *et al.*, 2005). Transformation of vegetation belts, for example, has been reported in the Spanish Central Range (Sanz-Elorza *et al.*, 2003), where high-mountain grassland communities dominated by *Festuca aragonensis*, typical of the cryo-Mediterranean belt, are being replaced by shrub patches of *Juniperus communis* ssp. *alpina* and *Cytisus oromediterraneus* from lower altitudes. Upward movements of species across the elevational gradient have also been reported in other Mediterranean mountain regions, such as the Montseny range in northeast Spain (Peñuelas *et al.*, 2007), although the reorganization of species' geographical ranges is likely the effect of different factors including climate change but also human management. Altitudinal changes in the distributions of butterfly species have also been reported in the Sierra de Guadarrama, central Spain (Wilson *et al.*, 2005), and, in the specific case of the *Parnassius apollo* butterfly, elevational changes in phenology influenced the temperatures experienced by larvae, and could also affect local host-plant favourability (Ashton *et al.*, 2009), thereby increasing the risk of this butterfly suffering population declines and local extinctions.

In the Apennines, bird species' distributions have changed in the last 20 years (Florenzano, 2004), and both diversity and composition of plant species have suffered significant transformations (Stanisci *et al.*, 2005). Climate warming may also imply drastic changes in the ranges of species, potentially leading to species inhabiting the upper belts to go extinct. Mountain conifer species such as *Pinus sylvestris*, *P. uncinata* and *Abies alba*, and other temperate species, especially *Fagus sylvatica* and *Quercus petraea*, were predicted to suffer a reduction in their range in the Iberian Peninsula (Benito-Garzón *et al.*, 2008). Specifically for the mountain conifer species, models predict reductions of 90% in the area of suitable climatic conditions under the most pessimistic scenarios for 2080. A recent study modelling distributions of European plants in relation to climate (Thuiller *et al.*, 2005) reported that mountain species, mainly those located near the Mediterranean Basin, were disproportionately sensitive to climate change (approximately 60% of species potentially lost by 2085), although these results should be interpreted cautiously due to the coarse resolution used. Similar results could be found for non-European Mediterranean mountains. *Cedrus atlantica* seems to be quite sensitive to climatic change, and modifications in the cedar's potential and real area of distribution will be considerable for 2100: models predict that *Cedrus atlantica* will find little refuge in the Atlas (Demarteau *et al.*, 2007).

Climate change is also affecting plant productivity. For example, growth of beech trees in the Apennines (Piovesan *et al.*, 2008) has declined in recent decades, in correspondence with increased drought. The patterns uncovered in that study suggest that long-term drought stress has reduced the productivity of beech forests in the central Apennines, in agreement with similar trends identified in other Mediterranean mountains. There are also changes in species phenology in relation to climate change, with changing dates in almost all the studied phenophases in recent decades (see Gordo and Sanz, 2009, for a recent study of plant phenology in Spain).

Many of the results reported above come from studies at local or regional scale. However, in the light of current global changes, assessing changes in biodiversity through time and across large spatial scales is one of the main challenges in the coming years for a better understanding of the biodiversity-climate relationships and also the impacts of climate change on ecosystem services. Studying ecosystem responses to changing climatic conditions could be accomplished using remote-sensing information and techniques (i.e. the Normalized Difference Vegetation Index, or NDVI; Figure 9.4). The remote-sensing information shows that over the last 20 years vegetation activity increased during spring and autumn but decreased during the summer months, suggesting an increase in inter-seasonal deviance (Figure 9.5 and Figure 9.6). These results also suggest that the time period for vegetation activity starts earlier in spring and finishes later in autumn than 20 years ago. Across the Mediterranean mountain ranges, the Atlas and Pyrenees encompass more areas with significant negative trends in vegetation activity, while the Apennines and Pindos are the ones less affected by decreases in vegetation activity (Figure 9.7).

Positive NDVI trends   **Negative NDVI trends**

Figure 9.4   Trends in vegetation activity in the 25 years from 1982 to 2006. Values of the Normalized Difference Vegetation Index (NDVI) were obtained from the Global Inventory Modelling and Mapping Studies (GIMMS). (*A full colour version of this figure appears in the colour plate section*)

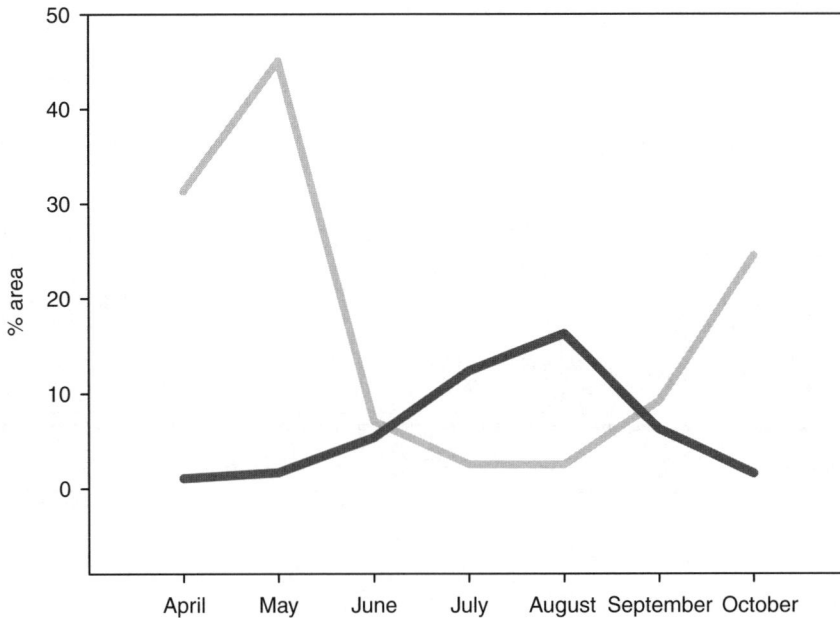

Figure 9.5   Average area across the six Mediterranean mountain ranges for significant positive (green) and negative (red) trends in the Normalized Difference Vegetation Index (NDVI) for the 25 years from 1982 to 2006

Climate change could affect mountain biodiversity directly but also through complex feedbacks with other vectors of global change. This complexity makes it more difficult to forecast future responses of biodiversity to climate change. For example, climate change significantly affects a key driver of Mediterranean ecosystems, namely fire. The increasing drought conditions in Mediterranean mountains predicted by climatic models may well increase the intensity and incidence of fires. Models predict vegetation shifts in the mountains of Greece that can be attributed to increased drought or to its associated changes in the local fire regime (Fyllas and Troumbis, 2009). Under the current climatic regime at Mount Olympus, models predict the future coexistence of *Pinus nigra* and *Quercus pubescens*. The interactions between species under climate change and their different dispersal/colonization abilities would imply a profound reorganization of ecological communities in Mediterranean mountains. Therefore we could expect that in future, novel and non-analogous communities will emerge and that species might be exposed to cascade extinction processes as a result of the disruption of biodiversity networks (Bascompte, 2009). Unfortunately, models predicting the complex future responses of biodiversity to climate change are still in their infancy. They may have, for example, underestimated declines for many specialist species, because species range contractions could increase due to a mismatch between the future distribution

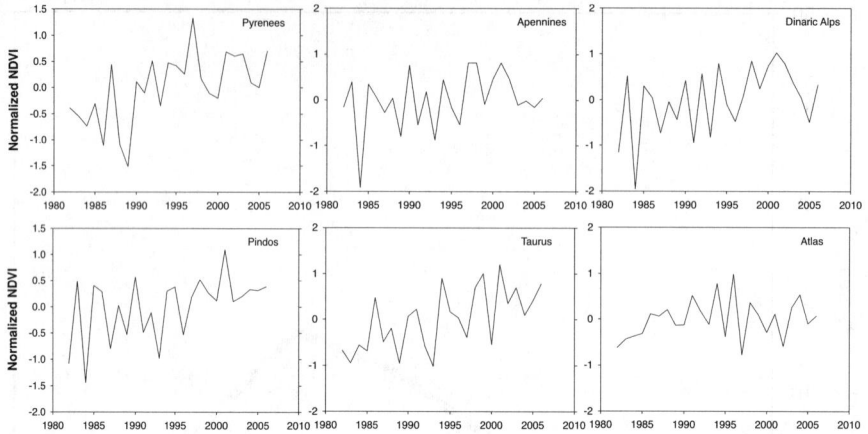

April

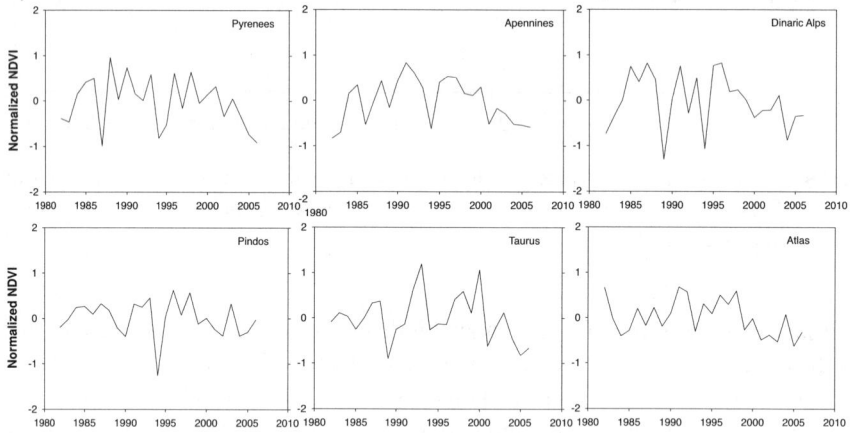

July

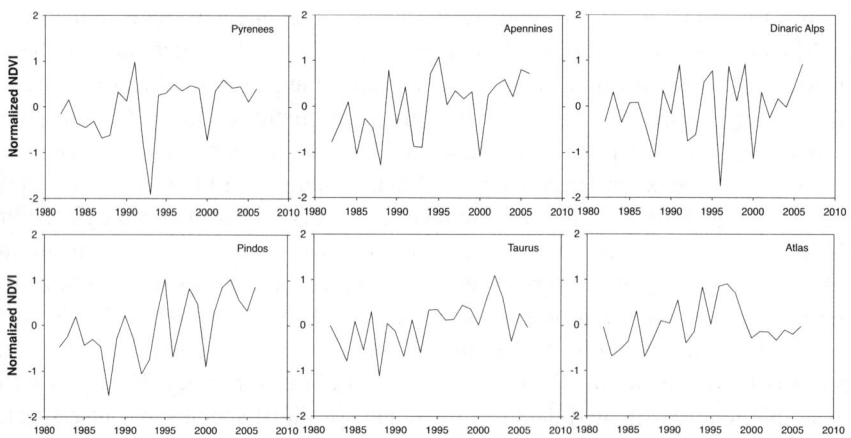

October

Figure 9.6 Trends in the Normalized Difference Vegetation Index (NDVI) from 1982 to 2006 for different mountain ranges in April, July and October

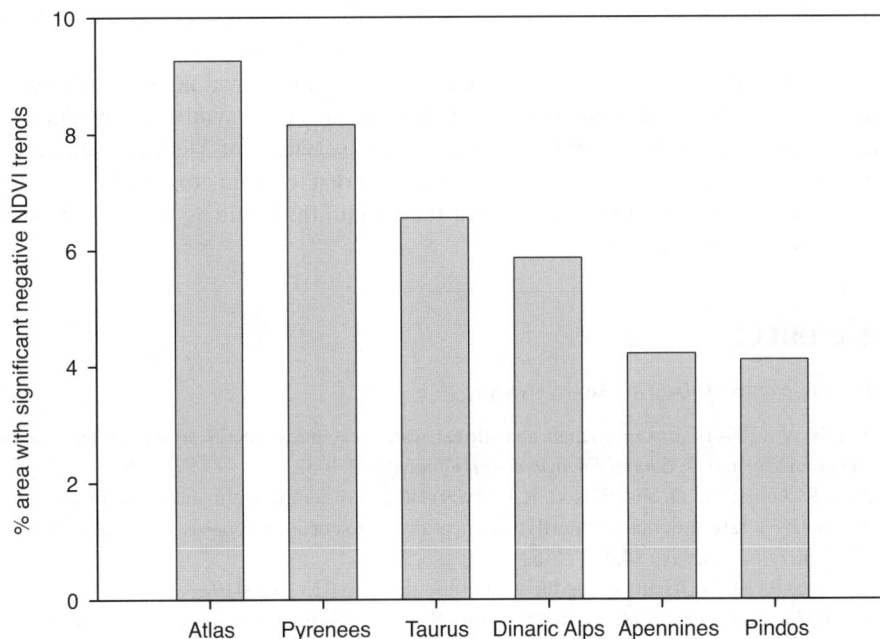

Figure 9.7   Area (% within each mountain range) with significant negative Normalized Difference Vegetation Index (NDVI) trends from 1982 to 2006. Values are an average of monthly values from April to October

of suitable climatic conditions and the availability of resources such as host plants (Merrill *et al.*, 2008). Specifically, Merril and colleagues show that climate is the most likely explanation for the low elevation-range margin of *Aporia crataegi* in the Sierra de Guadarrama (central Spain) whereas the absence of host plants from high elevations sets the upper limit. These examples aim to illustrate the complex relationships between biodiversity and climate change and the difficulty of predicting future climate change impacts on biodiversity, mainly in the highly biologically and topographically diverse Mediterranean mountains.

9.6   Conclusion

Mediterranean mountain environments seem to be accelerating towards uncertain ecological states because of changes associated with climate and land use changes (Nogués-Bravo *et al.*, 2008b). Models forecast significant increases in temperature and decreases in rainfall for the twenty-first century, and these changes will likely have an impact on (i) snow pack and glaciers, which also provide key habitats for alpine specialist species; and (ii) water availability and river discharge systems and therefore on aquatic and wetland habitats and species. Beyond these indirect effects on biodiversity, climate change would (iii) reduce habitat availability of alpine and subalpine belts increasing the risk of extinction for endemic species or range-restricted species, and may well disrupt the biological networks that ultimately support ecosystem functioning. Mountains have been symbols of wilderness and mystery but also the origins of resources and services for Mediterranean civilizations. Global and local strategic actions are needed to avoid crossing key thresholds in Mediterranean mountains so that they retain their role as key providers of services to human societies.

References

References marked as bold are key references.

Arnell, N.W. (2004) Climate change and global water resources: SRES emissions and socio-economic scenarios. *Global Environmental Change* 14:31–52.

Ashton, S., Gutierrez, D. and Wilson, R.J. (2009) Effects of temperature and elevation on habitat use by a rare mountain butterfly: implications for species responses to climate change. *Ecological Entomology* 34:437–446.

Bascompte, J. (2009) Disentangling the web of life. *Science* 325:416–419.

Beniston, M., Keller, F., Koffi, B. and Goyette, S. (2003) Estimates of snow accumulation and volume in the Swiss Alps under changing climatic conditions. *Theoretical and Applied Climatology* 76:125–140.

Benito-Garzón, M., Sanchez de Dios, R. and Sainz Ollero, H. (2008) Effects of climate change on the distribution of Iberian tree species. *Applied Vegetation Science* 11:169–178.

Bucher, A. and Dessens, J. (1991) Secular trends of surface temperatures at an elevated observatory in the Pyrenees. *Journal of Climate* 4:859–868.

Ceballos, A., Morán-Tejeda, E., Luengo-Ugidos, M.A. and Llorente-Pinto, J.M. (2008) Water resources and environmental change in a Mediterranean environment: The south-west sector of the Duero river basin (Spain). *Journal of Hydrology* 351:126–138.

Demarteau, M., Francois, L., Cheddadi, R. and Roche, E. (2007) Responses of *Cedrus atlantica* when faced with past and future climatic changes. *Geo-Eco-Trop* 31:105–146.

Diaz, H.F., Grosjean, M. and Graumlich, L. (2003) Climate variability and change in high elevation regions: past, present and future. *Climate Change* 59:1–4.

D'Orefice, M., Pecci, M., Smiraglia, C. and Ventura, R. (2000) Retreat of Mediterranean glaciers since the Little Ice Age: case study of Ghiacciaio del Calderone, Central Apennines. Italy. *Arctic, Antarctic and Alpine Research* 32:197–201.

Florenzano, G.T. (2004) Birds as indicators of recent environmental changes in the Apennines (Foreste Casentinesi National Park, central Italy). *Italian Journal of Zoology* 71:317–324.

Fujihara, Y., Tanaka, L., Watanabe, T., Nagano, T. and Kpjiri, T. (2008) Assessing the impacts of climate change on the water resources of the Seyhan River Basin in Turkey: Use of 45 dynamically downscaled data for hydrologic simulations. *Journal of Hydrology* 353:33–48.

Fyllas, N.M. and Troumbis, A.Y. (2009) Simulating vegetation shifts in north-eastern Mediterranean mountain forests under climatic change scenarios. *Global Ecology and Biogeography* 18:64–77.

Giorgi, F. (2002) Variability and trends of sub-continental scale surface climate in the twentieth century. *Climate Dynamics* 18:675–691.

Giorgi, F. (2006) Climate change hot-spots. *Geophysical Research Letters* 33:L08707.

Gordo, O. and Sanz, J.J. (2009) Long-term temporal changes of plant phenology in the Western Mediterranean. *Global Change Biology* 15:1930–1948.

Gritti, E.S., Smith, B. and Sykes, M.T. (2005) Vulnerability of Mediterranean Basin ecosystems to climate change and invasion by exotic plant species. *Journal of Biogeography* 33:145–157.

IPCC; Solomon, S. *et al.* (eds) (2007) *Contribution of Working Group I to the Fourth Assessment Report of the Intergovernmental Panel on Climate Change.* Cambridge, UK: Cambridge University Press.

Korner, C. (2004) Mountain biodiversity, its causes and function. *Ambio* (suppl.) 13:11–17.

López-Moreno, J.I. (2005) Recent variations of snowpack depth in the Central Spanish Pyrenees. *Arctic, Antarctic, and Alpine Research* 37:253–260.

López-Moreno, J.I. and García-Ruiz, J.M. (2004) Influence of snow accumulation and snowmelt on streamflow in the central Spanish Pyrenees. *Hydrological Sciences Journal* 49:787–802.

López-Moreno, J.I. and Vicente-Serrano, S.M. (2007) Mapping of snowpack distribution over large areas using GIS and interpolation techniques. *Climate Research* 33:257–270.

López-Moreno, J.I., Nogués-Bravo, D., Chueca-Cía, J. and Julián-Andrés, J. (2006) Change of topographic control on the extent of cirque glaciers since the Little Ice Age. *Geophysical Research Letters* 33:L24505.

López-Moreno, J.I., García-Ruiz, J.M. and Beniston, M. (2008) Environmental change and water management in the Pyrenees. Facts and future perspectives for Mediterranean mountains. *Global and Planetary Change* 66:300–312.

Manabe, S., Milly, P.C.D. and Wetherald, R. (2004) Simulated long-term changes in river discharge and soil moisture due to global warming. *Hydrological Sciences Journal* 49:625–642.

Marty, C. (2008) Regime shift of snow days in Switzerland. *Geophysical Research Letters* 35:L12501.

Médail, F. and Diadema, K. (2009) Glacial refugia influence plant diversity patterns in the Mediterranean Basin. *Journal of Biogeography* 36:1333–1345.

Merrill, R.M., Gutiérrez, D., Lewis, O.T., Gutiérrez, J, Díez, S.B. and Wilson R.J. (2008) Combined effects of climate and biotic interactions on the elevational range of a phytophagous insect. *Journal of Animal Ecology* 77:145–155.

Meybeck, M., Green, P. and Vörösmarty, C. (2001) A new typology for mountains and other relief classes: an application to global continental water resources and population distribution. *Mountain Research and Development* 21:34–45.

Myers, N. (2003) Biodiversity hotspots revisited. *BioScience* 53:916–917.

Nash, L.L. and Gleick, P.H. (1993) The Colorado River Basin and climatic change: The sensitivity of streamflow and water supply to variations in temperature and precipitation. EPA 230-R-93-009. Washington, DC: US Environmental Protection Agency.

Nogués-Bravo, D., Araújo, M.B., Errea, M.P. and Martínez-Rica, J.P. (2007) Exposure of global mountain systems to climate warming during the 21st century. *Global Environmental Change* 17:420–428.

Nogues-Bravo, D., Araújo, M., Lasanta, T. and López-Moreno, J.I. (2008a) Climate change in Mediterranean mountains during the 21st century. *AMBIO* 37:280–285.

Nogués-Bravo, D., Araujo, M.B., Romdal, T. and Rahbek, C. (2008b) Scale effects and human impact on the elevational species richness gradients. *Nature* 453:216–219.

Nunes, J.P., Seixas, J. and Pacheco, N.R. (2008) Vulnerability of water resources, vegetation productivity and soil erosion to climate change in Mediterranean watersheds. *Hydrological Processes* 22:3115–3134.

Oerlemans, J. (2005) Extracting a climate signal from 169 glacier records. *Science* 308:675–677.

Pandžić, K., Trninić, D., Likso, T. and Bošnjak, T. (2009) Long-term variations in water balance components for Croatia. *Theoretical and Applied Climatology* 95:9–51.

Parmesan, C. and Yohe, G. (2003) A globally coherent fingerprint of climate change impacts across natural systems. *Nature* 421:37–42.

Peñuelas, J., Ogaya, R., Boada, M. and Jump, A.S. (2007) Migration, invasion and decline: changes in recruitment and forest structure in a warming-linked shift of European beech forest in Catalonia (NE Spain). *Ecography* 30:829–837.

Piovesan, G., Biondi, F., Filippo, A.D., Alessandrini, A. and Maugeri, M. (2008) Drought-driven growth reduction in old beech (*Fagus sylvatica*) forests of the central Apennines, Italy. *Global Change Biology* 14:1265–1281.

Renard, B., Lang, M., Bois, P. *et al.* (2008) Regional methods for trend detection: Assessing field significance and regional consistency. *Water Resources Research* 44:W08419; doi:10.1029/2007WR006268.

Root, T.L., Price, J.T., Hall, K.R., Schneider, S.H., Rosenzweig, C. and Pounds, J.A. (2003) Fingerprints of global warming on wild animals and plants. *Nature* 421:57–60.

Sanz-Elorza, M., Dana, E.D., Gonzalez, A. and Sobrino, E. (2003) Changes in the high mountain vegetation of the central Iberian peninsula as a probable sign of global warming. *Annals of Botany* 92:273–280.

Stanisci, A., Pelino, G. and Blasi, C. (2005) Vascular plant diversity and climate change in the alpine belt of the central Apennines (Italy). *Biodiversity and Conservation* 14:1301–1318.

Thuiller, W., Lavorel, S., Araujo, M.B., Sykes, M.T. and Prentice, C. (2005) Climate change threats to plant diversity in Europe. *Proceedings of the National Academy of Sciences of the USA* 102:8245–8250.

Vicente-Serrano, S.M., Beguería, S. and López-Moreno, J.I. (2010) A multi-scalar drought index sensitive to global warming: The Standardized Precipitation Evapotranspiration Index – SPEI. *Journal of Climate* 23:1696–1718.

Viviroli, D. and Weingartner, R. (2004) The hydrological significance of mountains: from regional to global scale. *Hydrology and Earth System Sciences* 8:1016–1029.

Viviroli D., Dürr, H.H., Messerli, B., Meybeck, M. and Weingartner, R. (2007) Mountains of the world – water towers for humanity: typology, mapping and global significance. *Water Resources Research* 43:W07447.

Wilson, R.J., Gutierrez, D., Gutierrez, J., Martinez, D., Agudo, R. and Monserrat, V.J. (2005) Changes to the elevational limits and extent of species ranges associated with climate change. *Ecology Letters* 8:1138–1146.

Woodwell, G.M. (2004) Mountains: top down. *Ambio* (suppl.) 13:35–38.

Younger, P.L., Teutsch, G., Custodio, E., Elliot, T., Manzano, M. and Sauter, M. (2002) Assessment of the sensitivity to climate change of flow and natural water quality in four major carbonate aquifers of Europe. Geological Society Special Publication 193, pp. 303–323.

Zanchettin, D., Traverso, P., Tomasino, M. (2008) Po River discharges: A preliminary analysis of a 200-year time series. *Climatic Change* 89:411–433.

10
Conclusions

Ioannis N. Vogiatzakis

10.1 Introduction

Mountain ecosystems are important functional components of the planet and have provided model systems within the realms of ecological and sociocultural theories. Mediterranean mountains are typical examples of these systems, which display the diversity and complexity of the Mediterranean Basin itself. Although extensively terraced, grazed and planted, Mediterranean mountains are still considered natural and unspoilt. The rapid changes that are taking place in the Mediterranean (King *et al.*, 2001) are also affecting mountain environments in the area. These changes, some of which have been described in this book, are not only associated with the physical aspects but equally importantly with the socioeconomic responses to externalities. Mountains are influenced by a series of political, socioeconomic and cultural processes that take place within and beyond their boundaries. All Mediterranean mountains are similar in the following ways:

- The insular environment of mountains has a major impact on ecological and socioeconomic characteristics.

- Although biodiversity decreases with altitude, endemism increases, with mountains assuming the role of species refugia.

- They all provide a range of 'common/similar' ecosystem services (biodiversity, water, energy, traditional ecological knowledge) that extend beyond their boundaries.

- They have been and still are divine sanctuaries, have played a significant role in the development of the main religions in the region, and continue to provide refuge to mountain cultures.

- Human impact is quite limited compared to the lowlands, with agriculture, grazing and tourism being the main activities.

Mediterranean Mountain Environments, First Edition. Edited by Ioannis N. Vogiatzakis.
© 2012 John Wiley & Sons, Ltd. Published 2012 by John Wiley & Sons, Ltd.

- Future changes are uncertain due to climatic change, especially global warming, and its impact on a range of ecosystem services.

- Due to the harsh conditions particularly in the highest zones there is a limited range of options for sustainable development.

Despite their similarities there are a number of differences that make each mountain unique and give each a distinct character:

- Altitude variations resulting in differences in vegetation zonation; i.e. not all mountains present the same zonation from shrublands to alpine communities.

- Biogeographical elements that encompass Atlantic elements in the west, and Irano-Turanian in the east.

- The intensity and therefore impact of the principal human activities, i.e. agriculture, grazing and tourism, vary significantly from north to south of the basin.

- The major driver, i.e. population change, which is dependent on mountain services also varies between the northern and southern Mediterranean mountains.

- There is a significant increase in forest cover in the north of the Basin while in the south the trend is the opposite.

- There is a different history of colonization, invasion and exploitation by humans.

- Northern Mediterranean mountains are subject to EU laws and regulations regarding natural resources management and protection.

10.2 Environmental challenges in Mediterranean mountains

In a dynamic environment such as a mountain system, change is an inherent part of the processes that operate at a range of spatial and temporal scales in relation to both natural and socioeconomic factors. Mountain ecosystems and their associated services are the product of landscapes with a long history of land use, and therefore are susceptible to both climate change and socioeconomic transformations. There have been various reviews of the challenges that mountain regions face worldwide (see Beniston, 2003; Huber *et al.*, 2005; Borsdorf *et al.*, 2010). These challenges are inextricably linked with the provision of services (Körner and Ohsawa, 2005), which will be subject to pressures related to their demand and use, within and beyond the mountains' reach (Bayfield, 2001). The most important of these changes in the context of the Mediterranean mountains are discussed below.

10.2.1 Climate change

Although palaeoecological evidence for climate change exists for the Mediterranean, the sites are located in the lowlands, making extrapolation to mountains difficult. However, climate change is currently considered the driving force in ecosystem change during glacial-interglacial cycles (see Chapter 2). For the Mediterranean, models forecast significant increases in temperature and decreases in rainfall during the twenty-first century. These changes coupled with land use changes lead towards an uncertain future for the mountains of the Basin, as discussed in Chapter 9. Due to their close relationship with climate (especially temperature and precipitation), glaciers are the system most sensitive to climate change. Thus, the glacial record of the Mediterranean mountains has the potential to provide perhaps the clearest signal about past and future climate changes. However, as Hughes points out (see Chapter 3) there are still gaps in our knowledge of glacial geochronology in many mountains in the Mediterranean.

The amount and duration of rainfall in addition to the length of intervals between rainfall events are three import aspects of climate, the variation of which will influence geomorphological processes in mountain environments. Combined with changes in temperature these will result in changes in soil properties and accelerate desertification particularly in areas that border semi-arid areas (Geeson et al., 2002). An increase in the frequency of extreme events has been suggested as a direct result of climate change, manifesting itself as droughts and sediment transfer, and there is already evidence of that in Mediterranean mountain areas (Maas and Macklin, 2002).

There is evidence worldwide that climate change has already resulted in shifts in species ranges, timing of seasonal events and habitat preferences (Lenoir et al., 2008; Gottfried et al., 2012). In a mountain setting the effect on glaciers, snow regime and water regime will in turn have an impact on associated habitats and species. The possible reduction of habitat availability in the alpine and subalpine zone/belts increases the risk of extinction for endemic species or range-restricted species. The increase in frequency and intensity of fire outbreaks has been already reported from the Mediterranean areas (Mouillot et al., 2002; Pausas, 2004), and in the case of the Mediterranean mountains this is already occurring, as demonstrated by the large-scale fires in the mountains of the Peloponnese, south Greece, in August 2007. With the help of forest gap dynamics, in a case study in northeastern Mediterranean mountain forests, Fyllas and Troumbis (2009) showed that fire events should increase in number and will be associated with elevational shifts of the dominant tree species. Studies that have addressed the community-level responses to climate change in Mediterranean mountain ranges indicate that there will be colonization of high altitudes by subalpine species (Stanisci *et al.*, 2005; Kazakis *et al.*, 2007) or what Gottfried *et al.* (2012) termed the 'thermophilization' of mountain plant communities.

10.2.2 Socioeconomic challenges

Globally it is estimated that 13.3% of mountain land area is cultivated, while 50% is under some form of land use (Körner and Ohsawa, 2005). The Mediterranean Basin has historically been exploited with increasing intensity over time. Agriculture in particular has played a major role in shaping Mediterranean landscapes for the last 7000 years (Vogiatzakis *et al.*, 2006). This remains a major pressure, particularly in the south, where landscape degradation is mainly associated with population increase and the resulting human needs (Moore *et al.*, 1998). In the northern part of the Basin, land use has been driven by EU policies and directives, the most controversial perhaps of all being the Common Agricultural Policy, a major driver of landscape change in lowlands and uplands alike. At the same time new forms of land use (e.g. recreation, biodiversity protection, energy production, etc.) compete with traditional ones leading to conflicts.

By default perhaps, human land use in mountain areas promotes homogeneity and therefore reduces the inherent resilience of these systems. However, human activities in some Mediterranean mountains have resulted in land cover diversity (Nogués-Bravo, 2006). Price and Thompson (1997) argued that the dynamics of ecological and sociocultural systems in mountains are non-linear. In the mountains of the northern Mediterranean extensification of land use takes place, whereas in the south the trend is reversed, with more intense agricultural activity. There are also areas where little or no changes have been recorded in the last 50 years (Di Pasquale *et al.*, 2004; Hill *et al.*, 2004) partly because there have been limited social changes in these areas.

Mountain agriculture is practised on marginal lands, and where instruments have been formalized, as in the north of the Basin, agricultural activities have been supported. The abandonment of these activities has been detrimental to the equilibrium of these vulnerable systems, particularly in the north of the Basin, as demonstrated by Papanastasis in Chapter 8. Desertification is already an issue in the Mediterranean, therefore in areas where rainfall will diminish, further marginalization is expected (Geeson *et al.*, 2002).

Tourism in the Mediterranean has been one of the most important factors in the region's socioeconomic development in the last 50 years, resulting in broader socioeconomic and cultural changes. Tourism has been adopted universally as a tool for development, and in the case of the Mediterranean mountains has the potential to revive mountain economies. Worldwide, the mountain landscape itself is a main tourist attraction, which has resulted in a steady income to mountain communities but at the same time put more pressure on resources and cultural identities. In addition to tourists, local populations are now returning to the mountains, following a period of abandonment, mainly for cultural activities (see Chapter 7). Currently tourism as an external driver is non-uniform across the Mediterranean mountains. Again, its impact is concentrated mainly in the northern Mediterranean countries, which promote a range of tourism activities including mass tourism (e.g. ski resorts) but also ecotourism (e.g. trekking, agri-tourism, cultural tourism).

10.3 Adaptation and protection

10.3.1 Response to natural changes

Agenda 21 Chapter 13 of the Rio protocol ranks mountains among the most endangered landscapes worldwide due to their steep terrain and mountain climate in combination with intense land use pressure. This description recognizes on the one hand the potential for change as an inherent property of the mountain systems while on the other hand points out that human-induced disturbances will increase the impacts and decrease resilience. With the fast pace of natural and human-induced changes worldwide, adaptation has become a major topic of applied research in an attempt to restore the resilience of ecosystems but also prepare human communities for change.

Volcanic activity is one of the principal agents of natural disturbance. While this chapter was being written, Sicily's Mount Etna went through a series of eruptions from July to October 2011, causing lava sprays into the air but no damage or casualties. Since adaptation to volcanic activity is difficult, monitoring and early warning systems are probably the best option. Many of the authors in this volume have pointed out the need for increasing monitoring and development of early warning systems. For example De Jong *et al.* (in Chapter 5) advocate the analysis of snowmelt discharge regimes and their integration into seasonal forecasting as well as the development of adaptation strategies for mountain water resources. In some cases monitoring systems are in place, such as the GLORIA network (Global Observation Research Initiative in Alpine Environments; www.gloria.ac.at), an international initiative for assessing climate change impacts on mountain environments. Essentially it is the largest network worldwide for field data collection and monitoring of mountain climate and vegetation, which in the Mediterranean area includes the Sierra Nevada, Apennines, Lefka Ori (Greece) and the Corsican Alps.

Effort should be also directed towards modelling techniques, increasingly used to gain insight into the possible future changes in environments including mountains (see Chapter 9). Scale (both spatial and temporal), datasets and the technique employed are typical issues affecting the reliability of any modelling exercise. In addition, and since climate change will not act alone but in synergy with land use change, there is a need to decouple the effects of these factors and improve on the uncertainty of current predictions.

Modern nature conservation favours the adoption of a dynamic approach as a major response to climate change. Landscape approaches based on the use of corridors, buffer zones, ecological networks and greenways (Jongman and Pungetti, 2004) are increasingly gaining support (Bennett and Mulongoy, 2006; Hopkins *et al.,* 2007; Lawton, 2010). Based on the premise that landscape is the mosaic that encompasses agriculture, ecology and settlements, these approaches have the potential to safeguard ecosystems and their associated processes, as well as the traditional human practices with which these landscapes have evolved. The idea of protected landscapes, category V protected area designation according to the International Union for the Conservation of Nature (IUCN, 1994), was put forward in the early 1970s and included many mountain areas worldwide. Other initiatives that highlight the importance of protecting nature and culture under the umbrella of landscape have followed, such as the European Landscape Convention (Council of Europe, 2000).

10.3.2 Response to socioeconomic changes

Mountains in the political agenda

The year 2002 was declared International Year of Mountains, and since 2003 the 11 December has been designated by the UN Assembly as 'International Mountain Day', with the view to increase global awareness of the importance of mountains and promote them on the political agenda through the establishment of national committees in various countries. In addition, in Bangkok in November 2004, IUCN members approved resolution 3.039: 'The Mediterranean Mountains Partnership: . . . calling all concerned national, regional and local institutions to develop action plans for each of the major mountain ranges in the region, with the aim of achieving the conservation of their biological, landscape and cultural diversity, and boosting sustainable development' (Regato and Salman, 2008). These international efforts are not always accompanied by similar efforts at the national level, where often there is a lack of specific mountain policies.

Sustainability, the word of our times, implies a balance between economic development and environmental protection as a major response to change (cf. Morse, 2010). In mountain environments the development of relatively new economic resources such as tourism as well as the abandonment of rural communities and associated activities are perhaps the main socioeconomic challenges that need to be addressed. In order to tackle the disadvantages associated with mountain environments, often policies are introduced at the national or international level. One of the policy regulations heading towards that direction in the European Union is the designation of Least Favoured Areas (LFAs). Although currently under review, this mechanism has been in place since 1975. Recognizing the need for sustainable land management, the LFA scheme is part of the Rural Development Policy for 2007–2013 (EC, 2005), which aims at improving the environment and the countryside. As already discussed in this volume, agricultural activities in mountain environments are difficult due to natural handicaps. Agricultural land abandonment as a direct consequence of these adverse factors entails potential risks for mountain landscapes including biodiversity loss, desertification and forest fires. Therefore a payment scheme is introduced for these areas to mitigate these risks (EC, 2005). LFAs comprise, under Article 18 of Council Regulation 1257/1999, mountains among other areas: 'Mountain Areas are characterised as those areas handicapped by a short growing season because of a high altitude, or by steep slopes at a lower altitude, or by a combination of the two. Areas north of the 62nd parallel are also delimited as Mountains.'

The provision of services by mountain systems has been reiterated in this book and include biodiversity, water, energy and traditional ecological knowledge (MEA, 2005). The provision of, and more importantly the demand on, these ecosystem services extend beyond the mountain boundaries. Therefore finding ways to estimate these services in money terms, in a way that would benefit mountain communities, remains a challenge. Some examples of guidelines for Mediterranean mountain development are given in Table 10.1. Alternative (low-impact) tourism sector, participatory forestry and co-management schemes of protected areas are some of the pathways towards sustainable mountain development. In particular 'soft' tourism development may be related to reviving past activities through family businesses, which may lead to income generation while at the same time keeping culture alive (Benoit and Comeau, 2005). Of course it is difficult to predict the responses of human communities in mountain regions to the changing environment they live in (Funnell and Parish, 2001).

Table 10.1 Examples of guidelines for Mediterranean mountains' development (Regato and Salman, 2008)

Theme	Example guideline
1. Ecosystems management	Implementation of the Ecosystem Approach to support large-scale conservation and sustainable development
2. Watershed management	Maintenance and restoration of locally adapted management practices for mountain watersheds
3. Climate change adaptation	Design and implementation of participatory systems to monitor the effect of climate change
4. Mountain people	Enhance the self-esteem of mountain dwellers and correct the perception that society has of mountain cultures
5. Cultural and spiritual heritage	Use beliefs and attitudes held by mountain communities about the environment that could assist with the preservation of mountain integrity
6. Rural development	Increase funding opportunities/measures to restore the environment-development balance in mountains
7. Legal and institutional framework	Develop specific legal frameworks adapted to Mediterranean mountain societies
8. Transboundary cooperation	Develop a partnership framework between Mediterranean mountain areas

10.4   Conclusion

The ecological and cultural uniformity of the Mediterranean Basin has often been contested at the regional and local level, where diversification and complexity is evident (see Vogiatzakis *et al.*, 2008). Mountains are a typical example of this complexity as demonstrated in this volume, and decoupling the effects of the components contributing to their formation and subsequent shaping is an arduous task. What makes mountain systems unique is their complexity and scale. Therefore land use, management and conservation strategies should change in order to incorporate these elements. Although results of climate model predictions remain uncertain, climate change will be of overriding significance for future changes worldwide. What is certain is that climate change will not act alone but in conjunction with other stress factors, which will further threaten fragile ecosystems such as mountains.

Although monitoring and inventorying are on the increase in the Euro-Mediterranean countries, they are still limited for many other mountain areas in the Mediterranean (Kaya and Raynal, 2001). Even the knowledge of species distributions in many cases is poor and incomplete (Vogiatzakis *et al.*, 2003; Tvrtkovic and Veen, 2006). What is yet to be carried out is a concerted and coordinated effort to account for all the services provided by Mediterranean mountains. Despite the fact that they are often used as political boundaries between countries, mountain massifs in physical terms represent a single entity and therefore should be managed as such. Often the state of effort and knowledge is dissimilar on either side of a mountain massif. In addition there is a need for undertaking Environment Impact Assessments where development projects affect Protected Areas, and to ensure their recommendations are enforced and monitored. Generally in the Mediterranean there is inadequacy of legislation or ineffective enforcement as well as a lack of political commitment to sustainable development, although there is a divide between countries of the northern and southern Mediterranean (Vogiatzakis *et al.*, 2006).

Future research needs to be coordinated in an interdisciplinary and international manner to optimize the identification of particularly vulnerable regions and provide a portfolio of adaptation and mitigation options. Since responses to change, whether ecological, institutional or social, are non-linear there needs to be a better research model and decision-making framework for Mediterranean mountains. In the words of the late Zev Naveh (1987) '. . . a holistic landscape-oriented ecological approach supporting integrated and dynamic conservation and multi-purpose management'.

References

References marked as bold are key references.

Bayfield. N. (2001) Mountain resources and conservation. In: Warren, A. and French, J.R. (eds), *Habitat Conservation Managing the Physical Environment*. John Wiley & Sons, Ltd.
Beniston, M. (2000) *Environmental Change in Mountains and Uplands*. London: Arnold.

Beniston, M. (2003) Climatic change in mountain regions: A review of possible impacts. *Climatic Change* 59:5–31.

Bennett, G. and Mulongoy, K. (2006) *Review of Experience with Ecological Networks, Corridors and Buffer Zones*. Montreal: Secretariat of the Convention on Biological Diversity.

Benoit, G. and Comeau, A. (2005) *A Sustainable Future for the Mediterranean: The Blue Plan's Environment and Development Outlook*. Earthscan.

Borsdorf, A., Grabherr, G., Heinrich, K., Scott, B. and Stötter, J. (eds) (2010) *Challenges for Mountain Regions – Tackling Complexity*. Vienna: Böhlau Verlag.

Council of Europe (2000) *European Landscape Convention*. European Treaty Series No. 176. Council of Europe.

Di Pasquale, G., Garfi, G. and Migliozzi, A. (2004) Landscape dynamics in south-eastern Sicily in the last 150 years: the case of the Iblei Mountains. In: Mazzoleni, S., di Pasquale, G., Mulligan, M., di Martino, P. and Rego, F. (eds), *Recent Dynamics of the Mediterranean Vegetation and Landscape*. Chichester: John Wiley & Sons, Ltd, pp. 73–80.

EC (European Council) (2005) Council Regulation (EC) No. 1698/2005 of 20 September 2005 on support for rural development by the European Agricultural Fund for Rural Development (EAFRD).

Funnell, D.C. and Parish, R. (2001) *Mountain Environments and Communities*. **London: Routledge.**

Fyllas, N.M. and Troumbis, A.Y. (2009) Simulating vegetation shifts in north-eastern Mediterranean mountain forests under climatic change scenario. *Global Ecology and Biogeography* 18:64–77.

Geeson, N., Brandt, C.J. and Thornes, J.B. (2002) *Mediterranean Desertification: a Mosaic of Processes and Responses*. John Wiley & Sons, Ltd.

Gottfried, M., Pauli, H., Futschik, A. *et al.* **(2012) Continent-wide response of mountain vegetation to climate change.** *Nature Climate Change* **2:111–115.**

Hill, J., Hostert, P. and Roder, A. (2004) Long-term observations of Mediterranean ecosystems with satellite remote sensing. In: Mazzoleni, S., di Pasquale, G., Mulligan, M., di Martino, P. and Rego, F. (eds), *Recent Dynamics of the Mediterranean Vegetation and Landscape*. Chichester: John Wiley & Sons, Ltd, pp. 33–43.

Hopkins, J.J., Allison, H.M., Walmsley, C.A., Gaywood, M. and Thurgate, G. (2007) *Conserving Biodiversity in a Changing Climate: Guidance on Building Capacity to Adapt*. London: DEFRA.

Huber, U.M., Bugmann, H.K.M. and Reasoner, M.A. (2005) *Global Change and Mountain Regions: an Overview of Current Knowledge*. **Springer.**

IUCN (1994) *Guidelines for Protected Area Management Categories*. CNPPA with the assistance of WCMC. Gland, Switzerland, and Cambridge, UK: International Union for the Conservation of Nature.

Jongman, R. and Pungetti, G. (eds) (2004) *Ecological Corridors and Greenways: Concepts, Design, Implementation*. Cambridge: Cambridge University Press.

Kaya, Z. and Raynal, D.J. (2001) Biodiversity and conservation of Turkish forests. *Biological Conservation* 97:131–141.

Kazakis, G., Ghosn, D., Vogiatzakis, I.N. and Papanastasis, V.P. (2007) Vascular plant diversity and climate change in the alpine zone of the Lefka Ori, Crete. *Biodiversity and Conservation* 16:1603–1615.

King, R., De Mas, P. and Beck, J.M. (eds) (2001) *Geography Environment and Development in the Mediterranean*. Sussex Academic Press.

Körner, C. and Ohsawa, M. (coords. lead authors) (2005) Mountain systems. In: Hassan, R., Scholes, R. and Ash, N. (eds) *Millennium Ecosystem Assessment*. Island Press, Chapter 24.

Lawton, J.H., Brotherton, P.N.M., Brown, V.K. *et al.* (2010) *Making Space for Nature: a Review of England's Wildlife Sites and Ecological Network*. Report to Defra.

Lenoir, J., Gegout, J.C., Marquet, P.A. *et al.* (2008) A significant upward shift in plant species optimum elevation during the 20th century. *Science* 320:1768–1771.

Maas, G.S. and Macklin, M.G. (2002) The impact of recent climate change on flooding and sediment supply within a Mediterranean mountain catchment, southwestern Crete, Greece. *Earth Surface Processes and Landforms* 27:1087–1105.

MEA (Millennium Ecosystem Assessment) (2005) *Ecosystems and Human Well Being*. Island Press.

Moore, H.M., Fox, H.R., Harrouni, M.C. and El Alami, A. (1998) Environmental challenges in the Rif Mountains, northern Morocco. *Environmental Conservation* 25:354–365.

Morse, S. (2010) *Sustainability: A Biological Perspective*. Cambridge University Press.

Mouillot, F., Rambal, S. and Joffre, R. (2002) Simulating climate change impacts on fire frequency and vegetation dynamics in a Mediterranean-type ecosystem. *Global Change Biology* 8:423–437.

Naveh, Z. (1987) Landscape ecology, management and conservation in the European and Levant Mediterranean uplands. In: Tenhunen, J.D., Catarino, F.M., Lange, O.L. and Oechel, W.C. (eds), *Plant responses to Stress: Functional Analysis in Mediterranean Ecosystems*. Springer-Verlag, pp. 641–657.

Nogués-Bravo, D. (2006) Assessing the effect of environmental and anthropogenic factors on land-cover diversity in a Mediterranean mountain environment. *Area* 38:432–444.

Pausas, J. (2004) Changes in fire and climate in the eastern Iberian peninsula (Mediterranean Basin). *Climatic Change* 63:337–350.

Price, M. and Thompson, M. (1997) The complex life: human land uses in mountain ecosystems. *Global Ecology and Biogeography Letters* 6:77–90.

Regato, P. and Salman, R. (2008) *Mediterranean Mountains in a Changing World; Guidelines for Developing Action Plans*. IUCN.

Stanisci, A., Pelino, G. and Blasi, C. (2005) Vascular plant diversity and climate change in the alpine belt of the central Apennines (Italy). *Biodiversity and Conservation* 14:1301–1318.

Tvrtkovic, N. and Veen, P. (eds) (2006) *The Dinaric Alps Rare Habitats and Species: A Nature Conservation Project in Croatia*. Zagreb: Croatian Natural History Museum and Royal Dutch Society for Nature Conservation (KNNV).

Vogiatzakis, I.N., Griffiths, G.H. and Mannion, A.M. (2003) Environmental factors and vegetation composition Lefka Ori massif, Crete, S. Aegean. *Global Ecology and Biogeography* 12:131–146.

Vogiatzakis, I.N., Mannion, A.M. and Griffiths, G.H. (2006) Mediterranean ecosystems: problems and tools for conservation. *Progress in Physical Geography* 30:175–200.

Vogiatzakis, I.N., Pungetti, G. and Mannion, A. (eds) (2008) *Mediterranean Island Landscapes: Natural and Cultural Approaches*. Landscape Series Vol. 9. Springer Publishing.

Index

Mediterranean Mountain Environments, First Edition. Edited by Ioannis N. Vogiatzakis.
© 2012 John Wiley & Sons, Ltd. Published 2012 by John Wiley & Sons, Ltd.